William Ewart

Cardiac Outlines for Clinical Clerks and Practitioners

and first principles in the physical examination of the heart for the beginner

William Ewart

Cardiac Outlines for Clinical Clerks and Practitioners
and first principles in the physical examination of the heart for the beginner

ISBN/EAN: 9783337272494

Printed in Europe, USA, Canada, Australia, Japan

Cover: Foto ©berggeist007 / pixelio.de

More available books at **www.hansebooks.com**

FOR CLINICAL CLERKS
AND PRACTITIONERS

AND

FIRST PRINCIPLES IN THE PHYSICAL EXAMINATION
OF THE HEART FOR THE BEGINNER

BY

WILLIAM EWART, M.D. Cantab., F.R.C.P.

PHYSICIAN TO ST. GEORGE'S HOSPITAL, CLINICAL LECTURER AND TEACHER OF PRACTICAL
MEDICINE IN THE MEDICAL SCHOOL, PHYSICIAN TO THE BELGRAVE HOSPITAL FOR
CHILDREN, ADDITIONAL EXAMINER IN 1891 FOR THE THIRD M.B. OF THE
UNIVERSITY OF CAMBRIDGE, LATE ASSISTANT PHYSICIAN AND
PATHOLOGIST TO THE BROMPTON HOSPITAL FOR CON-
SUMPTION AND DISEASES OF THE CHEST

WITH SIXTY-TWO ILLUSTRATIONS

G. P. PUTNAM'S SONS

NEW YORK   LONDON
27 WEST TWENTY-THIRD STREET   24 BEDFORD STREET, STRAND
The Knickerbocker Press
1892

Electrotyped, Printed, and Bound by
The Knickerbocker Press, New York
G. P. Putnam's Sons

PREFACE.

The publication of this collection of Cardiac Outlines was undertaken with a threefold object : to help the beginner through his first difficulties, to encourage the clinical clerk in the cultivation of the graphic method as a means to thoroughness and accuracy, and to place at the disposal of the practitioner an easy method of adequately recording important clinical observations. This little book does not profess to be a treatise on cardiac diseases. With the exception of short and incidental pathological sketches it is entirely devoted to the study of physical examination under healthy and morbid conditions. My aim has been to provide the reader with a method which he may find equal to the requirements of cardiac study, and which may help him onwards farther than I am yet competent to lead.

It would have been impossible to have set forth the method as a whole without including some details not strictly elementary. At the same time the needs of the junior student have been steadily considered, both in the beginning which has been made rudimentary, and in the gradual progression to more advanced subjects.

I am under a pleasing obligation to my nephew, Mr. P. de Vaumas, for valuable assistance in the production of several of the illustrations.

WM. EWART,
33 Curzon St., May Fair,
London, W.

New York, March, 1892.

CONTENTS.

vii

PART II

THE DERMOGRAPHIC METHOD AND THE CONSTRUCTIVE SERIES OF OUTLINES.

PART III.

THE PRACTICAL METHODS OF INSPECTION AND PALPATION.

PART IV.

CARDIAC PERCUSSION AND THE "PERCUSSION" SERIES OF OUTLINES (NORMAL AND PATHOLOGICAL).

Contents.

PART V.

CARDIAC AUSCULTATION IN THE NORMAL SUBJECT AND IN PATHOLOGICAL CONDITIONS.

Contents.

PART VI.

PRACTICAL ILLUSTRATIONS OF THE METHOD OF USING DIAGRAMS AND
SYMBOLS FOR THE RECORD OF PHYSICAL EXAMINATIONS OF THE
HEART :

INTRODUCTORY REMARKS.

ON THE VALUE OF ANATOMICAL PRECISION AND ON THE USE OF THE GRAPHIC METHOD IN THE CLINICAL STUDY OF THE HEART.

Most teachers will probably agree with the author in tracing to an imperfect appreciation of anatomical facts a large share of the trouble usually experienced in the earlier stages of clinical study ; and in regarding the graphic method as the surest means of safeguarding the beginner against the evils of an inadequate anatomical training.

Medical anatomy, so far as it relates to the heart, presents no intrinsic difficulties. It commonly happens, however, that the student is already cut off from the opportunities of dissecting when a knowledge of visceral anatomy becomes indispensable to him. Too often he is apt to regard medical anatomy as distinct from anatomy in the strict sense. Any lack of anatomical accuracy in the clinical manuals placed in his hands confirms this impression ; and his anatomical notions and his clinical studies are kept asunder at the time when their combination would have been of most use.

Clinical anatomy, or the anatomy of the living organism, is in reality not less precise than the anatomy of the dead body ; on the contrary, it exacts a knowledge of additional detail. We should be imperfectly guided in our clinical work and in our visceral surgery were we to trust alone to the relations of parts such as they are learned in the dissecting-room. The main anatomical facts remain as a basis, but on this groundwork a knowledge of the configuration of the living viscera has to be elaborated. This is the special office of percussion and of auscultation. In the hands of the expert these methods may be said to take up and to complete the work of dissection.

On the other hand it is of obvious advantage to the inexperienced percussor that he should be guided by broad anatomical facts. Beginners would make slow progress in percussion and auscultation without the help

of anatomy ; and in percussion of the heart, which is probably more difficult and more important than that of most other organs, this assistance is
more needful than elsewhere.

In the arrangement of the work the author has been guided by a belief
that anatomical knowledge should be made the starting-point, whilst it is
also the goal of the clinical methods of percussion and auscultation. We
should learn to percuss with the help of anatomical data ; whilst the highest
office of percussion is to enable us to trace the individual anatomy of each
chest with an accuracy not attainable by any other means.

The graphic method affords us welcome aid in both stages of this scheme.
The anatomical facts essential for a clinical study of the heart can be most
quickly taught, and with greatest advantage, by means of a graduated series
of diagrams, which the student should learn to trace for himself. This practice will confer not only a knowledge of normal anatomy, but a facility for
drawing diagrams of abnormal hearts.

Independently of the educational purpose in view, a special need for
some anatomical detail arose in these pages in connection with some of the
principles of cardiac percussion, which will probably be regarded as novel,
and might, in the absence of anatomical proof, have been deemed arbitrary.

For the Outlines themselves no artistic value can be claimed. They
are simply diagrams, and in many respects imperfect ; but special care has
been taken to avoid any glaring departure from anatomical principles on
all essential points. Exception might be taken to the dimensions and to
the shape of the thorax. The preference given to the broad type of
chest was guided by a practical purpose, and the somewhat unusually large
size will probably be found a convenience in note-taking. Again the
manifest unreality of a simultaneous tracing of the ribs and of the heart has
its justification in the attempt to facilitate a localisation of the heart and of
its sounds. Diagrams constructed on this plan may be difficult to decipher
at first, but if any one of them be understood the others will not puzzle the
reader.

THE DIVISION OF THE WORK.

The Outlines are divided into the following groups :

(1) A descriptive or anatomical series gives a sketch of the thorax, of the
    liver, of the heart, and of the pericardium.

(2) A " constructive " series illustrates the way in which a diagram of the heart can be correctly traced on the chest of the living or of the dead subject.

(3) A few pages are devoted to the methods of physical examination and to inspection and palpation in health and in disease.

(4) Percussion of the heart under normal and under pathological conditions is dealt with in the next series of Outlines.

(5) The section devoted to cardiac auscultation is preceded by remarks on the normal and pathological heart sounds, and by codes of symbols for use in describing abnormal heart sounds and cardiac murmurs by means of the graphic method.

(6) Lastly, examples are given showing the manner of using the diagrams and symbols for clinical records.

ELEMENTARY CLINICAL OUTLINE OF THE CHEST.

A simple diagram of the front of the chest, useful for clinical purposes, can be constructed with six lines and a long vertical arrow.

The head of the arrow stands for the *xiphoid cartilage ;*

The notch at the other end for the *episternal notch.*

The lines represent :

> The two *clavicles,*
> The two *costal arches,* and
> The *lateral outline* of the thorax.—
> The *nipples,* and
> The *umbilicus*

complete the essential parts of the diagram ; the dotted lines not being essential.

This rough diagram can be rapidly sketched, requires no knowledge of drawing, and will therefore enable any observer to add to his notes a fairly accurate graphic record of the physical signs discovered in the chest.

FIG. I.

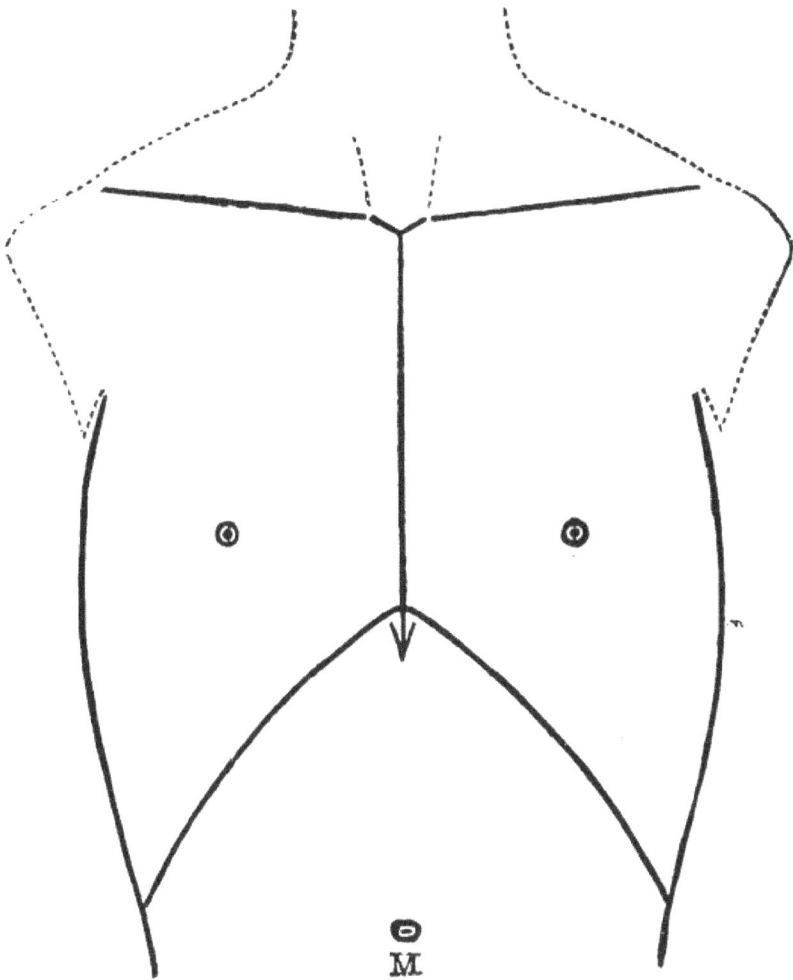

The lines represent : the clavicles, the sternum, the costal arch, the nipples, and the umbilicus (M).

5

THE IMAGINARY LINES OF REFERENCE.

The following lines only are used as lines of reference in connection with the heart :

> The middle line,
> The vertical nipple line,
> The parasternal line.

THE MIDDLE LINE.

The thorax is very often far from symmetrical ; and in order to judge of the extent of the asymmetry, and for other reasons, we require a middle line which shall be true.

How to Trace a True Line on the Chest.—Stretch a cord between the two points selected. Along this a pencil or marking chalk may be drawn. Better still is the plan suggested to me by a valued pupil, Mr. L. Moysey. The cord or thread, having been previously made damp with a little ink, is lifted off the skin between the nails of two fingers and allowed to fly back like the string of a bow, the result being a perfectly straight and distinct ink-mark.

How to Determine the Middle Line.—We select with this view the umbilicus, which remains almost always strictly median in position, and the middle of the episternal notch. This point also is generally in the middle line, so long as no great thoracic deformity exists, and that the clavicles are of even length (neither fractured nor unusually bent). By uniting these two points a line will be obtained which very frequently will pass to one side or the other of the tip of the sternum. Deviations of this bone, and therefore of the long axis of the thorax, being much less considerable above than below, the episternal notch is usually a reliable guide.*

---

* In special cases of deformity it may be necessary to strike a line between the middle line of the face, and the umbilicus (or even the symphysis pubis), whilst the patient is standing perfectly straight.

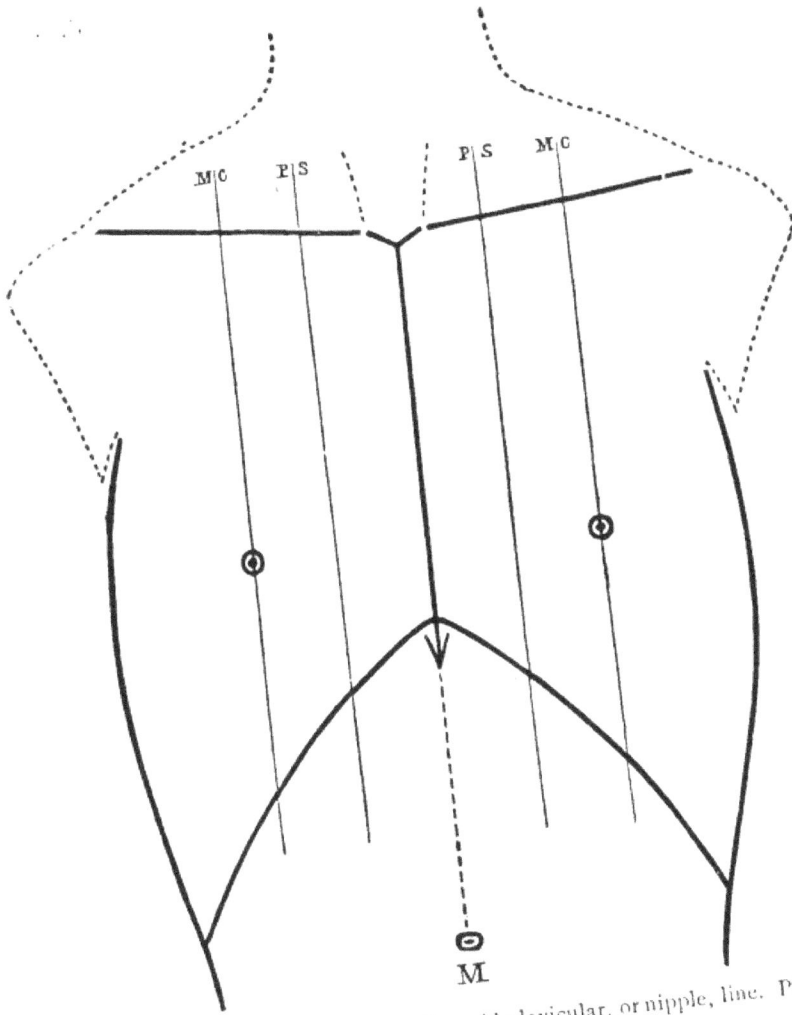

FIG. II.

M—The umbilicus. MC—The mid-clavicular, or nipple, line. PS—The parasternal line.

THE VERTICAL NIPPLE LINE.

The nipples being very liable to vary in site, and in the adult female being moreover mobile, are not perfect landmarks, although constantly used as such. *A vertical line drawn through the middle of the clavicle* coincides with the nipple line, when this is normally situated. The mid-clavicular line has the advantage of starting from a fixed spot, and is therefore to be preferred ; it is indicated in the diagram under the letters MC.

THE PARASTERNAL LINE.

This line is of great practical importance. It lies half way between the other two. The parasternal line, as well as the mid-clavicular, being parallel to the middle line, can be accurately drawn with the help of one instead of two landmarks.

PART I.

THE DESCRIPTIVE OR ANATOMICAL SERIES OF OUTLINES.

The Outlines belonging to this series are intended to illustrate points in the anatomy of the heart and of its surroundings, possessing clinical interest or importance.

THE THORAX.

NOTICE in this Outline :

(1) The width of the manubrium ;
(2) The rigid sternal attachment of the first rib ;
(3) The overlapping of the clavicle, leaving no space above the first rib ;
(4) The transverse ridges intersecting the sternum ;
(5) The upper line of junction (double dotted) between the manubrium and the gladiolus. This junction remains cartilaginous up to a considerable age or permanently, and capable of bending at an angle, which is then known as the Angulus Ludovici.

ANGULUS LUDOVICI.

This is a rather frequent deformity dependent upon the cartilaginous nature of the junction just described. If, under the forces at work, the sternum be bent forwards (Angulus Ludovici), the chest space will be enlarged by so much. If it be bent backwards (other form of Angulus Ludovici), chest space will be lost.

Angular deformity of any other part of the sternum is seldom seen. The xiphoid joint, however, is very pliable, and the xiphoid cartilage subject to various displacements.

THE COSTAL CARTILAGES.

(1) The 1st chondrocostal junction lies external to the oblique line joining the others.

(2) The 4th chondrocostal junction lies about an inch internal to the nipple.

(3) The 1st and 2d cartilages slant upwards from the sternum ; the 3d is usually horizontal ; the others leave the sternum at an angle, downwards.

(4) All, except the 1st, are connected with the sternum by articulations.

(5) The 6th and 7th cartilages articulate with each other so as to strengthen the costal arch.

(6) *The two 7th cartilages form the costal arch*, in conjunction with the end of the gladiolus ; they receive the insertion of the 8th cartilages.

(7) The 8th cartilages receive the insertion of the 9th ; and the 9th that of the 10th.

(8) The 11th cartilages may be felt at the sides ; they are free from any connection with the costal arch. The 12th cartilages cannot be seen or felt from the front.

FIG. III.

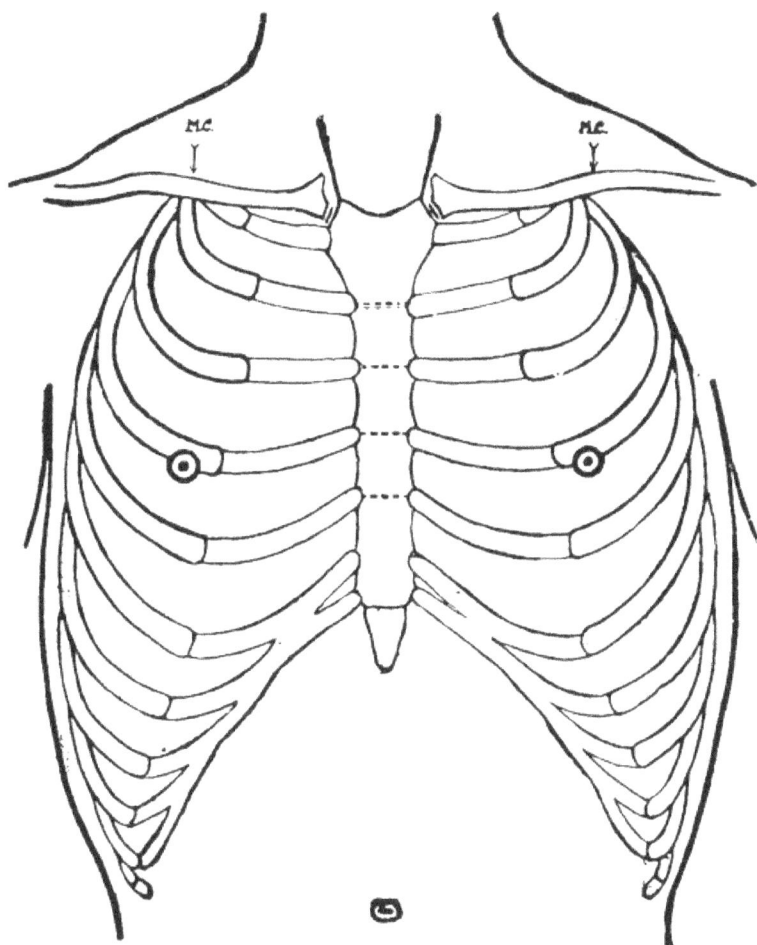

The direction of the mid-clavicular line (MC) is indicated by the arrows.
N. B.— *The lowest sternal segment is too long in the diagram.*

11

THE LIVER, STOMACH, AND SPLEEN IN THEIR MUTUAL RELATION AND IN THEIR RELA-TION TO THE THORAX.

The *diaphragm* is not shewn in the Outline; nevertheless its position may be readily gathered, its convex surface following the line SH and the direction of the same line continued on the left side down to S; whilst the line of attachment of the membrane coincides in the main with that of the costal arch.

The stomach, although not normally in immediate contact with the heart, is separated from it only by a small thickness of liver.

The spleen is seen to be far removed from the region X of the cardiac beat.

THE LIVER IN SITU.

A knowledge of the position of the liver is essential to our description of the heart, since the liver may be said to form the heart's basis of support.

The following points will be noticed in the diagram :

(1) The general resemblance of the outline of the liver (*in situ*) to that of a cocked hat, one extremity pointing straight to the left, the other extremity straight down ;

(2) Its almost complete inclusion within the thorax, with the exception of

(3) A small portion left unprotected by cartilage or bone at the epi-gastrium.

(4) It occupies the right two thirds of the dome of the diaphragm, the left third of which shelters the stomach and spleen.

(5) Its thin and sharp inferior and anterior edge (IH in the figure) is in contact with the anterior thoracic wall along the line of the costal arch ; and the same contact is kept up by the anterior zone of the convex surface of the liver, for some small distance upwards.

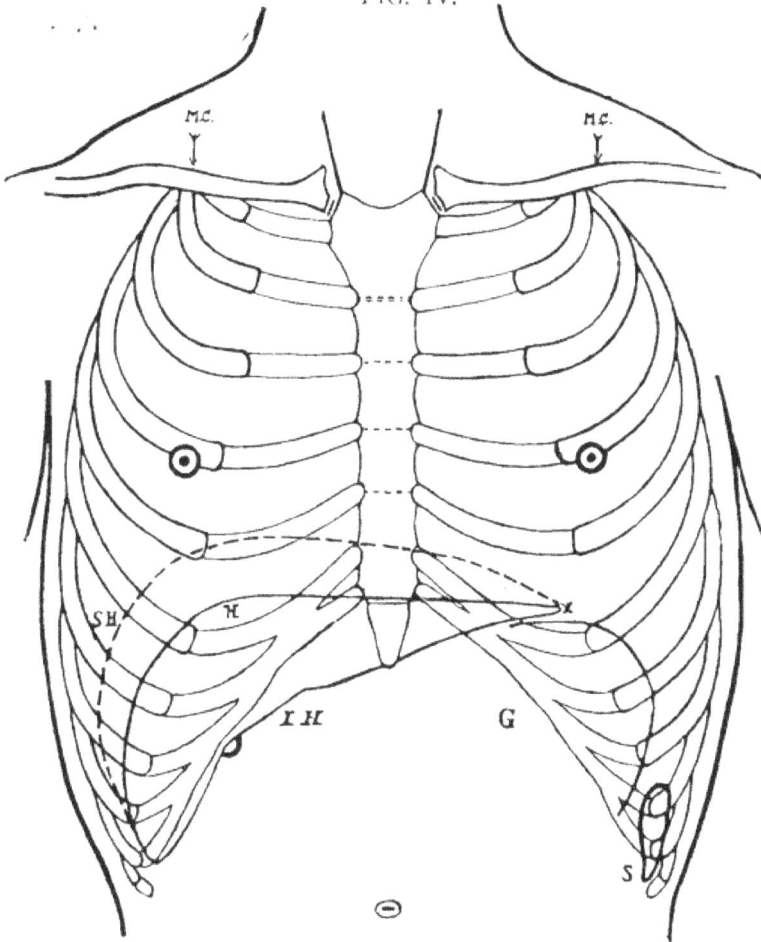

FIG. IV.

G—The semilunar line of gastric resonance. S—The spleen. SH—
The upper boundary of the liver within the thorax (suprahepatic line). III
—The lower border of the liver. H—The upper hepatic boundary in direct
contact with the anterior chest-wall (hepatic line). X—The left extremity
of the liver.

N. B.—*The left lobe of the liver should have extended slightly beyond the
left nipple line in the diagram.*

(6) The line along which the contact between the diaphragm covering the liver and the anterior thoracic wall ceases is, for convenience, termed in these pages the *Hepatic Line* (H in the figure).

(7) Above this level the convex surface still rises considerably ; but it recedes as it rises.

(8) The inferior surface, owing to the way in which the bulk of the liver finds room in the depth of the thorax, reveals in a front view little of the great thickness of the posterior hepatic border ; on the contrary, this surface, which is slightly concave, slants upwards and backwards, as viewed from below.

(10) The right extremity is thick.

(11) The left extremity is very thin ; it extends farther than depicted in the Outline, and at least as far as the mid-clavicular line.

N. B.—The hepatic line of thoracic contact, H, has a very slight fall towards the left, which is not made sufficiently evident in the diagram.

THE HEPATIC BOUNDARIES.

These all vary during deep respiration. During tranquil respiration the displacements are moderate, and the average levels would be as follows :

The *Inferior Hepatic Line*, IH, nearly coincides with the right costal arch as far as the tip of the 8th rib ; thence crossing the epigastrium as a tangent to the tip of the xiphoid (this relation varies much) it rises to the left extremity (X) of the organ.

The *Hepatic Line*, H, passes nearly horizontally through the sterno-xiphoid joint, having a slight inclination downwards towards the left. In the right hypochondrium it is convex to the right, and follows a downward course between the 6th and 10th chondrocostal junctions.

The *Supra-hepatic Line*, SH, is almost horizontal, at a level just below that of the right 5th chondrocostal junction ; in its course both to the left and downwards it is almost parallel with the hepatic line just described. Its left extremity is slightly curved downwards.

The Fundus of the Gall-bladder lies to the inner side of the tip of the right 9th rib.

The boundary between the right and the left lobes (at III) is situated nearly half way between the tip of the 9th rib and that of the xiphoid cartilage.

N. B. The hepatic boundaries will be again alluded to in Part IV., and will form the subject of systematic study. Reference will also be made to the changes in level to which these boundaries are liable as a result of the respiratory variations in the capacity and in the contents of the chest.

THE PRÆCORDIUM AND THE PRÆVASCULAR AREA.

The *Præcordium*, or præcordial region, is, in the broader sense, that part of the anterior thoracic surface which covers the heart (see also Outline VI). It includes :

(1) The sternum from the upper level of the 3d cartilages to the xiphoid ;

(2) The *left* 4th and 5th cartilages from rib to sternum, and the 4th interspace ;

(3) The inner third of the *left* 3d and 6th cartilages, and the sternal extremity of the 7th ; also the greater part of the 3d and of the 5th intercartilaginous spaces ;

(4) The inner third of the *right* 4th, 5th, and 6th cartilages, and of the 3d, 4th, and 5th intercartilaginous spaces.

It is best not to apply the term præcordium to the more limited area to be described in Part IV. as the area of absolute dulness, over which the cardiac surface is covered only by the thoracic wall.

The *Prævascular Area* is of much smaller extent. It is continued upwards from the præcordium, as far as the lower clavicular level, and corresponds to the upper segment of the gladiolus, to the manubrium, and to the inner third of the 1st and 2d cartilages and of their interspaces.

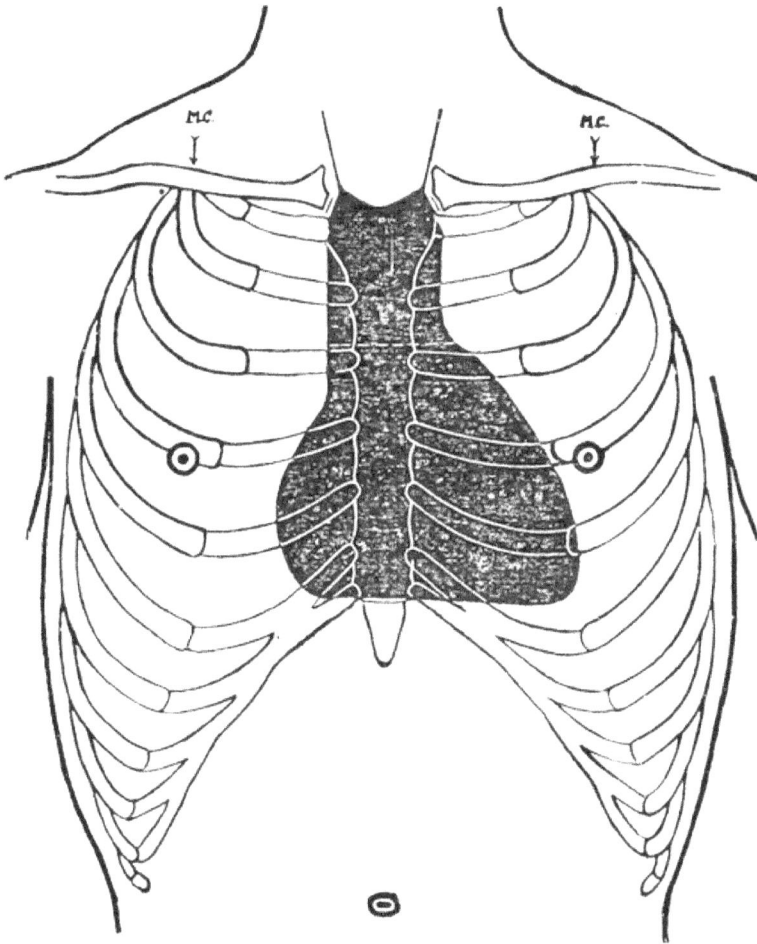

FIG. V.

The surface above the interrupted line is the prævascular area.   The surface below the same line is the præcordial area.

N. B.—*The " base of the heart" does not strictly correspond with the line given in this diagram (cf. page 26, and Outline XI.).*

THE PERICARDIAL SAC AND MEMBRANE.

No cavity (in the sense of free space) exists in the pericardium in health : besides the heart, the pericardium contains only a few drops of fluid.

The *Sac* is capacious in comparison with the size of the heart, and fits the latter loosely (especially at the sides).

The *Membrane* is described as possessing a *visceral* and a *parietal* layer.

The *visceral* layer covers the entire free cardiac surface, and accompanies the great vessels for a short distance. Being reflected from them, it is continued into the parietal layer.

The *parietal* layer forms the serous sac. It is loose at the sides only, being elsewhere *adherent* to various structures ; the most important of these are :

> The œsophagus,
> The great vessels, and
> The central tendon of the diaphragm.

It is adherent to :

> The thymus and the sternum in front,
> The œsophagus and tracheal bifurcation behind,
> The great vessels and the deep fascia of the neck above, and
> The central part of the diaphragm below.

The heart therefore possesses some freedom of vertical movement and considerable freedom of lateral movement ; but its antero-posterior range of movement is limited.

THE PLEURO-PERICARDIUM.

This is a convenient anatomical and clinical name for the loose lateral portion of the parietal layer, which is intimately adherent to, and forms one membrane with, the corresponding loose portion of the parietal layer of the pleura. This double membrane has thus a pleural face and a pericardial (or mediastinal) face, and is a mobile partition between the pericardial cavity and the pleural cavity. It is attached to the root of the lung and encloses the phrenic nerve.

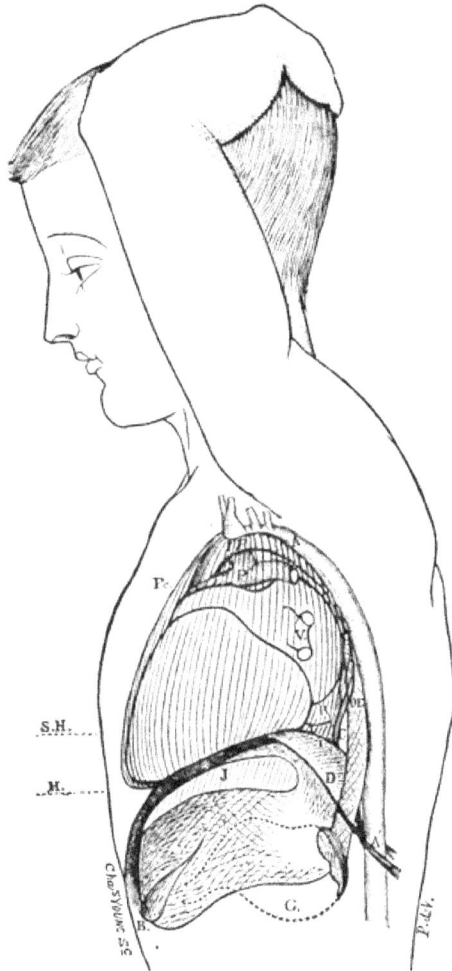

FIG. VI.

A—The aorta. B—The gall-bladder. D—Diaphragm. Œ—The œsophagus. J—The left hepatic lobe, in vertical section. G— The fundus of the stomach, in section. H—The anterior hepatic level (hepatic line). SH—The superior hepatic level (suprahepatic line). Pc—The pericardial membrane.

19

THE BOUNDARIES OF THE FLOOR OF THE PERICARDIUM.

(1) It is seen that the left half of the chest contains more of this surface than the right (almost in the proportion of $\frac{4}{5}$ to $\frac{1}{5}$, or at least of $\frac{2}{3}$ to $\frac{1}{3}$), and that the portion included in the right half broadens from front to back.

(2) The surface of the floor of the pericardium is polygonal or very irregularly circular, with some resemblance to a lozenge : it includes the surface of the middle or anterior division of the Centrum tendineum, and of some of the anterior muscular bundles of the diaphragm, more particularly on the left side.

(3) It occupies the greater portion of the interval between the base of the xiphoid appendix and the œsophagus.

(4) The right border is oblique outwards and backwards, forming an anterior angle with the end of the sternum, and a posterior angle at the orifice of the vena cava. The posterior border extends from the latter to the left, in front of the œsophagus. The left border is convex outwards and almost semicircular between the œsophagus and the left 7th chondrosternal joint, with increased convexity at the apex site, X. The anterior border is very short, nearly corresponding to the line of the sterno-xiphoid joint.

(5) The outline of the floor of the pericardium might therefore be described as *polygonal rather than circular, but without any sharp angles.*

(6) Its longest diameter extending from left front to right back.

(7) The outline further displays the situation of the orifice of the Vena Cava, presently to be described, and of the phrenic nerves and accompanying vessels.

FIG. VII.

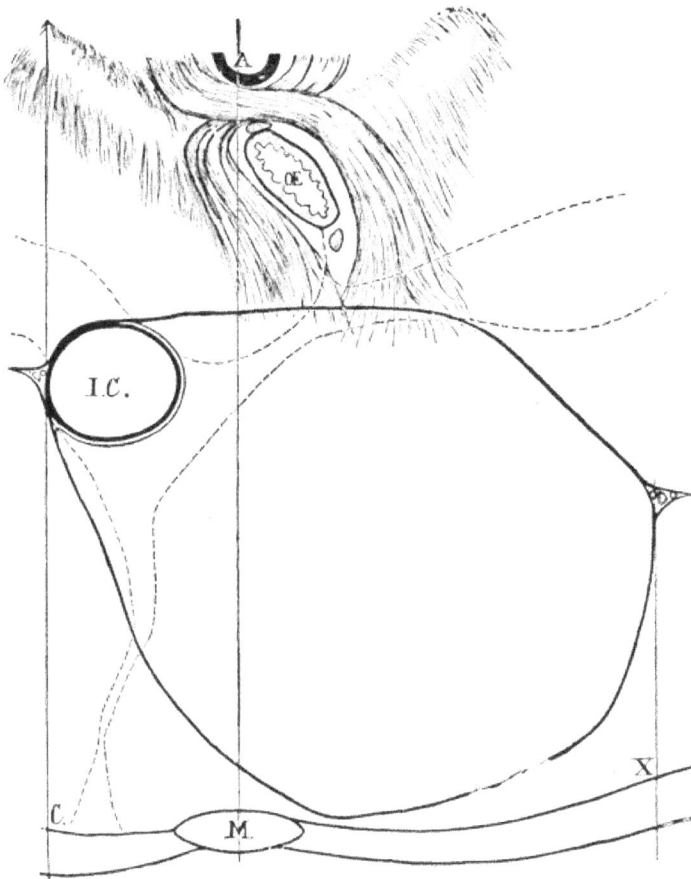

M—Middle line of body, and tip of sternum, attached to which are the costal cartilages. A—Aorta encircled by crura of diaphragm. Œ—Œsophagus and pneumogastric nerves. IC—Inferior vena cava. X—Extreme left boundary line (line of cardiac apex). C—Extreme right boundary line (line of vena cava).

N. B.—*The interrupted lines have reference to the ligaments of the liver.*

THE INCLINATION OF THE PERICARDIAL FLOOR.

This Outline displays several of the points described in Outline VII. But in addition it gives some idea of the inclination of the floor of the pericardium. This is inclined on the horizontal in two directions :

(1) A very slight inclination from the right to the left, downwards ;

(2) A very marked inclination from back to front, downwards.

The resulting inclination might be readily demonstrated by tilting a small oblong four-legged table. If the two legs on one of the long sides (that farthest from the observer) be made longer than the other two, so that the table slopes towards the observer ; and if the two (now uneven) legs on the observer's left be raised to an equal but very small extent, by placing under them a thick board, the lowest spot of the whole surface will be at the anterior corner facing the observer's right. This is precisely the slope of the pericardial floor.

On the contrary, the highest spot of the table's surface would obviously be the right posterior corner (observer's left). On referring to the two diagrams of the pericardial floor it will be seen that this spot is occupied by the orifice of the Inferior Vena Cava. It is important therefore to remember that the orifice of the Inferior Vena Cava into the right auricle is

(1) The highest and (2) the extreme right spot

at the heart's lower surface ; it being understood that the heart is in direct contact with this part of the floor of the pericardium.

On the other hand, to the heart's apex is allotted

(1) The lowermost and (2) the extreme left position.

The way in which the serous layer of the membrane is reflected from the side and floor of the pericardium to the left surface of the vein at its termination is shown in Outline VII. Both Outlines show that the parietal layer, on the contrary, is applied and adherent to the right surface of the vein. By this means the corresponding part of the right auricle is bound down to this spot of the pericardial floor ; and this is the only attachment which the lower surface of the heart suffers.

The Outline also shews the Œsophagus in its close relation to the posterior wall of the pericardium ; and the Inferior Vena Cava receiving the capacious Hepatic Veins.

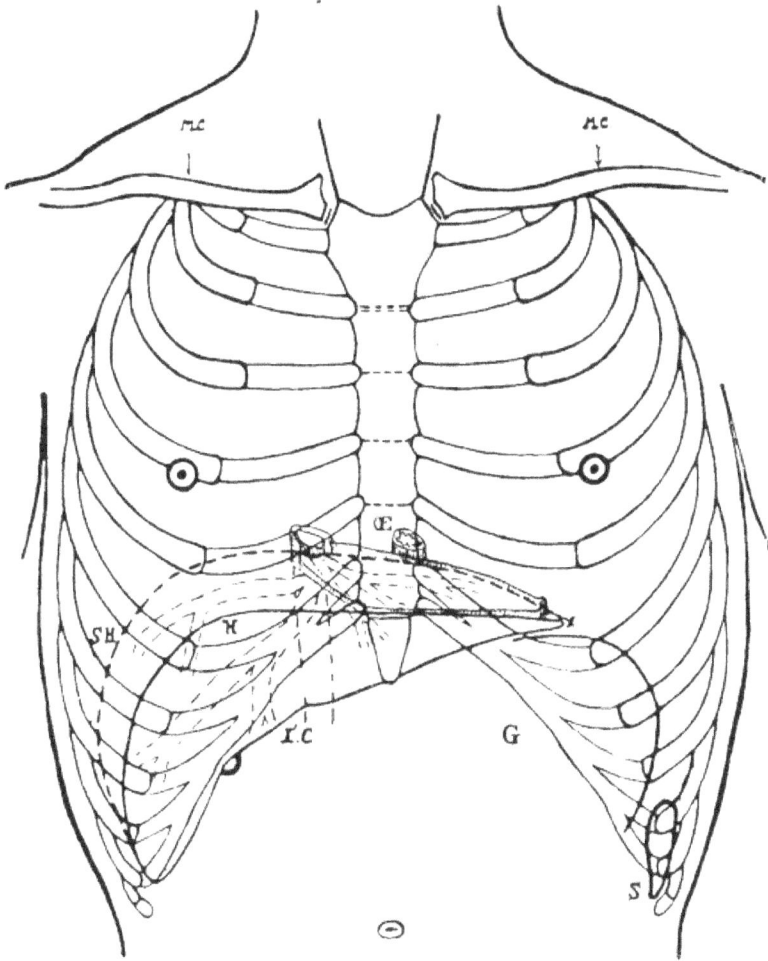

FIG. VIII.

ŒE.—The œsophagus (behind the pericardium). IC—The inferior vena cava, terminating just within the pericardium. H—The anterior hepatic level, and the anterior attachment of the pericardium. SH (horizontal portion of SH)—The posterior hepatic level, and the posterior attachment of the pericardium.

THE SHAPE AND POSITION OF THE HEART AND ITS RELATION TO THE LIVER AND RIBS.

The anterior surface of the heart is here exposed to view. This displays :

(1) The entire width and height of the Right Ventricle ;

(2) A small portion of the Left Ventricle ;

(3) The anterior surface of the Right Auricle ;

(4) A portion of the Left Auricular Appendix ;

(5) The origin of the Aorta, Pulmonary Artery, and Superior Vena Cava ;

(6) (The outline of the Left Auricle would not be seen from the front, in the normal subject ;)

(7) The Apex of the Right Ventricle and that of the Left ;

(8) The " Base " of the heart :

(9) The obliquity of the heart's longitudinal axis downwards, forwards, and to the left :

(10) The anterior position of the Right cavities and the deep position of the Left cavities ;

(11) The lower level of the Right cavities ;

(12) The higher level of the Left cavities (which indirectly results from the convexity of the upper surface of the liver) ; and

(13) The relatively high level of the orifice of the Inferior Vena Cava, due to the same cause ;

(14) The relations of the anterior cardiac surface to the thorax, to the nipple, to the liver, and to the stomach and spleen.

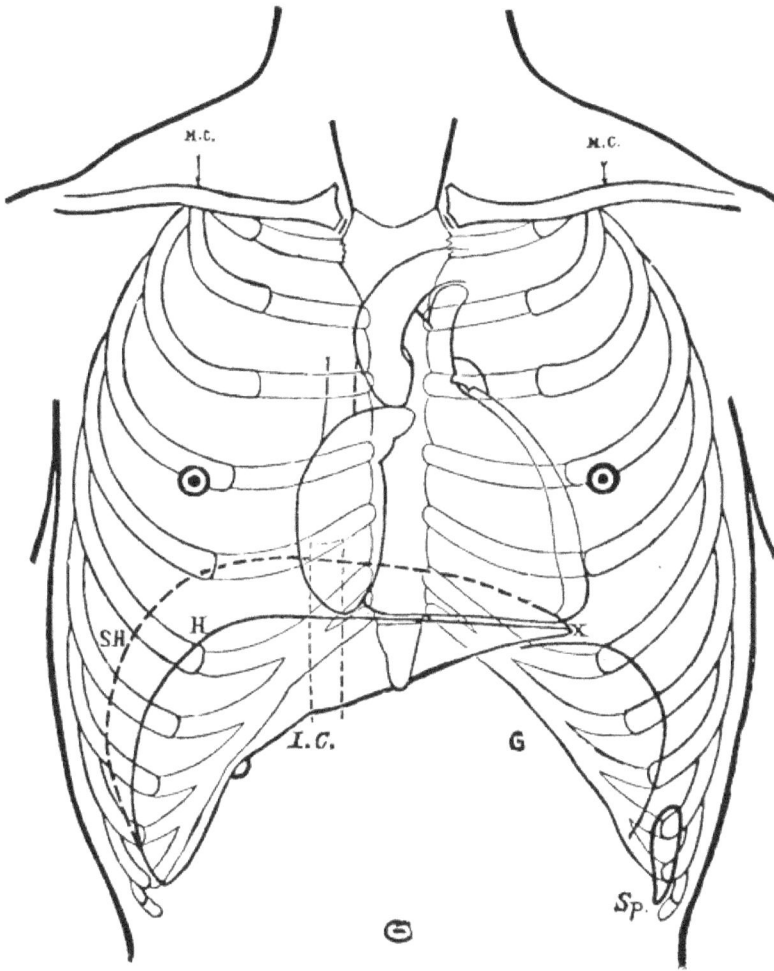

FIG. IX.

The letters refer to the same structures as in previous Outlines.

25

THE CARDIAC APEX AND BASE.

THE APEX.

The heart's apex is not a mere point but a rounded surface, formed chiefly by the left ventricle. On its left aspect it is covered by lung ; but during systole it moves slightly from under cover towards the right. That portion of the cardiac surface which is felt beating, and is commonly known as the "apex-beat," is the *right aspect* of the apex ; it belongs in part to the right ventricular wall.

THE POSITION OF THE CARDIAC APEX.

The normal position of the heart cannot be described in terms of measurement, since the size of individuals varies ; but it is correct to localize the average site of the apex in the 5th left interspace, a little internal to the mid-clavicular line, and nearer to the 6th than to the 5th rib.

THE POSITION OF THE BASE OF THE HEART.

The present Outline, viewed apart from any preconceived notions, would suggest that the heart resembles a cone or a pyramid. A line XS (see Outline XI.) passing through the cardiac apex, X, and through the top of the right shoulder, might be regarded as the axis of the heart as well as the axis of the Septum Ventriculorum : and the base of the heart would be perpendicular to this line.

The origin of the Aorta and that of the Pulmonary Artery being situated at the upper extremity of the line XS, are correctly described as belonging to the cardiac base. At the same time the reader will presently gather from Outline XI. that, strictly speaking, the plane of the base of the heart would be included between the planes LL' (of the base of the left ventricle) and RR' (of the base of the right ventricle), and would therefore be represented in the diagram by a line much more oblique than that which runs through the orifices of the Aorta and of the Pulmonary Artery.

Clinically the term "base" is applied somewhat indefinitely to the region including the sternal end of the right and left 3d cartilages and of the 2d interspaces. It would probably be variously defined by different authorities. In this uncertainty the student should bear in mind the following points :

(1) The plane of the cardiac base is not correctly represented by a horizontal line such as is shown, for instance, in Fig. VI.

(2) Its extent is not correctly represented by the distance which intervenes between the "aortic" or right 2d interspace, and the "pulmonary" or left 2d interspace.

(3) On the contrary, the plane of the cardiac base *has an oblique direction and it diverges from the horizontal line joining the 2d interspaces ;*

(4) For, whilst the left 2d interspace corresponds to the origin of the Pulmonary Artery, the right 2d interspace does not correspond to the origin of the Aorta.

(5) Therefore, although the second sound is listened for at the right 2d interspace, this is not the site of its production ;—and similarly, although in clinical language we may refer to murmurs audible there as "basic" murmurs, this should not lead us to suppose that the 2d interspace coincides in position with the plane of the base of the heart.

THE RIGHT VENTRICLE.

The anterior wall of the Right Ventricle and of the Pulmonary Artery has been removed, exposing to view their cavities.*

The position of the Tricuspid Valve and of the Semilunar Valves is roughly indicated.   Specially to be noticed are :

(1) The Conus Arteriosus, seen in vertical section, as it crosses the root of the Aorta which lies behind it ;

(2) The Septum between the ventricles, S, bulging towards the right ;

(3) The Lower or Right flap of the Tricuspid Valve ;

(4) The Posterior Pulmonary Semilunar flap, PP, immediately covering the origin of the Aorta (the other two flaps are anterior).

(5) Notice that the anterior surface (see Outline IX.) being roughly triangular, and the inferior or diaphragmatic and the septal surfaces being also triangular, the cavity of the Right Ventricle represents a pyramidal space.

THE RIGHT AURICLE.

The Right Auricle occupies the space between the right border of the sternum and the extreme right cardiac boundary close to the parasternal line.   Some idea of its shape and size may be formed by remembering that it extends backwards as far as the orifice of the Inferior Vena Cava ; and that its highest point anteriorly is formed by the tip of the Right auricular appendix, behind the middle of the sternum, at the level of the 3d rib.

* In this and the three following Outlines the tracing of the ribs has been interrupted over the præcordium, to avoid confusion.

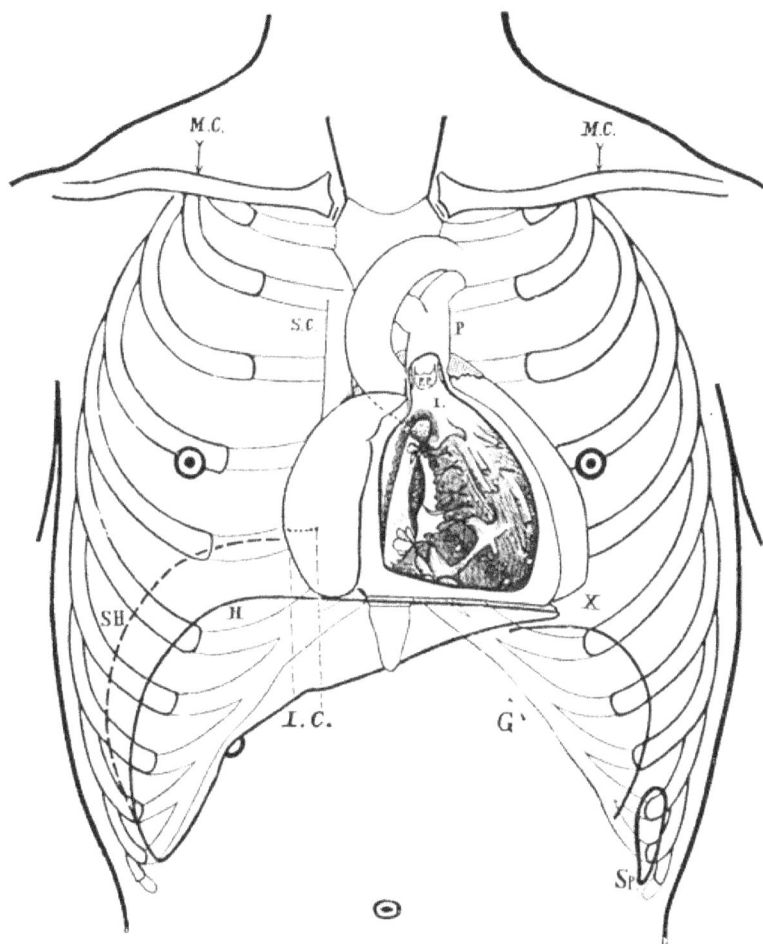

SC—The superior vena cava. IC—The inferior vena cava. I—The infundibulum laid open from the front, together with the right ventricle. P—The pulmonary artery ; behind which the shaded surface of the left auricle. PP—The posterior pulmonary valve flap ; the two anterior flaps have been removed with the front wall of the vessel.

THE SEPTUM VENTRICULORUM, THE CARDIAC AXIS, AND THE CARDIAC BASE.

In the erect posture the Septum forms an almost vertical partition, XS, between the two ventricles. Its course is not absolutely straight, but slightly curved with convexity towards the right ventricle. Its plane is neither strictly sagittal nor transverse ; but it passes *obliquely* from the apex, backwards and to the right, and slightly upwards, towards the spine. The following points can be made out in the Outline :

(1) With its lower border the Septum rests upon the diaphragm.

(2) This lower border, or foot, is therefore not horizontal, but, like the upper surface of the liver, it slopes upwards and backwards, nearly parallel to the line XS.

(3) The anterior or upper border is superficial, and corresponds to the anterior interventricular groove. (See Outlines IX. and X.—In the present Outline the upper part of the septum is supposed to have been cut away.)

(4) At its anterior extremity the Septum forms the apex of the heart ; it possesses here no appreciable height. From this point backwards it quickly becomes taller, acquiring its greatest height halfway between the apex and the base of the ventricles.

(5) It assists in the formation of the Conus Arteriosus, or Infundibulum, of which it may be regarded as the left and inferior wall.

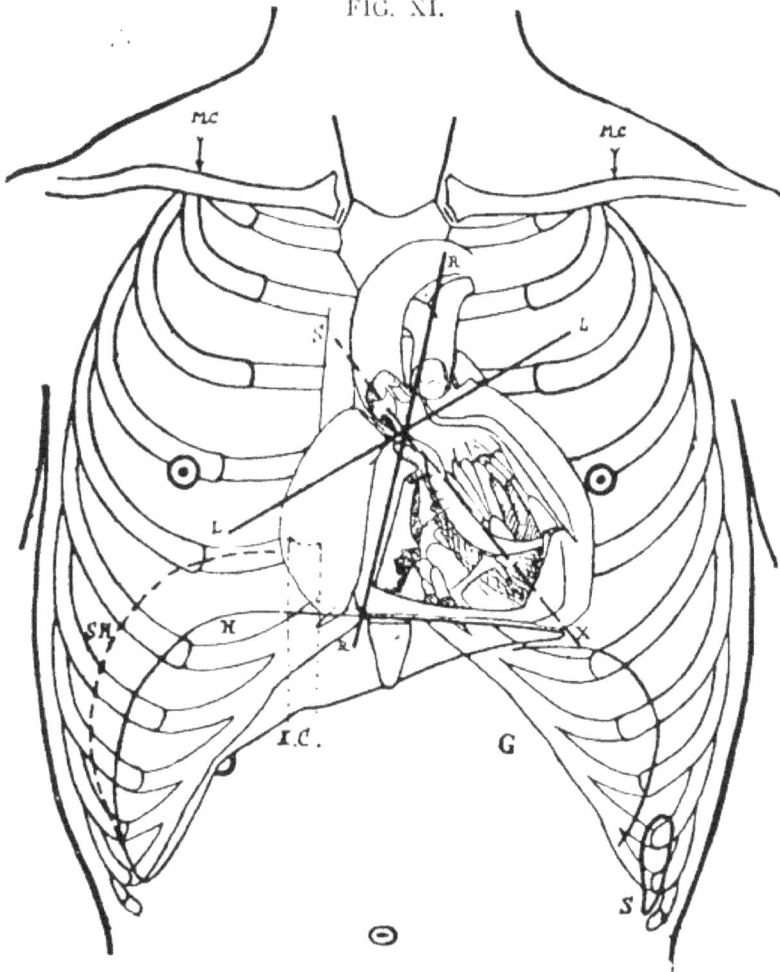

FIG. XI.

XS—Axis of the heart, passing through the septum ; the septum having been partly removed, a view is obtained of the left as well as of the right ventricular cavity. LL'—Line passing through the plane of the base of the left ventricle. RR'—Line passing through the plane of the base of the right ventricle.

N. B.—*The plane of the base of the heart as a whole would fall between these two planes.*

31

THE LEFT VENTRICLE.

This Outline is intended to show the relation which the cavity of the Left Ventricle bears to that of the Right Ventricle, and to the Septum. The Septum, exposed by removing the anterior wall of the right ventricle as in the previous Outline, is supposed to have been partly cut away, and with it also a portion of the anterior wall of the left ventricle.

The same incisions carried upwards have excised (between letters P and l) that portion of the conus arteriosus and pulmonary artery which lies in front of the origin of the aorta. A view is thus obtained (to the left of the Septum, S) of the interior of the left ventricle, and of the beginning of the aorta.

With the help of these explanations, the reader will easily recognise the following structures :

(1) The Aorta, A, laid open, showing the Semilunar Valves, namely, the Right posterior, part of the Left posterior, and part of the Anterior flap.

(2) The Pulmonary Artery, P, divided and partly removed.

(3) The Conus Arteriosus, I, partly removed.

(4) The foot of the Septum, S, across which lie some of the divided chordæ tendineæ of the Tricuspid valve.

(5) The Mitral valve, M, connected by means of its chordæ tendineæ with the Musculi Papillares below.

(6) The left lateral situation of the Septum causes the left ventricle to be entirely contained within the left half of the chest.

(7) *A fortiori* the Mitral valve, which occupies the left posterior and upper corner of the ventricle, is well to the left of the middle line, and of the sternum.

(8) The upper part of the Septum being separated from the surface by the thickness of the Conus Arteriosus which lies in front of it, we realize the still deeper position taken by the Left ventricular cavity.

(9) It follows that the Aortic valve is situated rather deeply behind the left side of the sternum ; and that

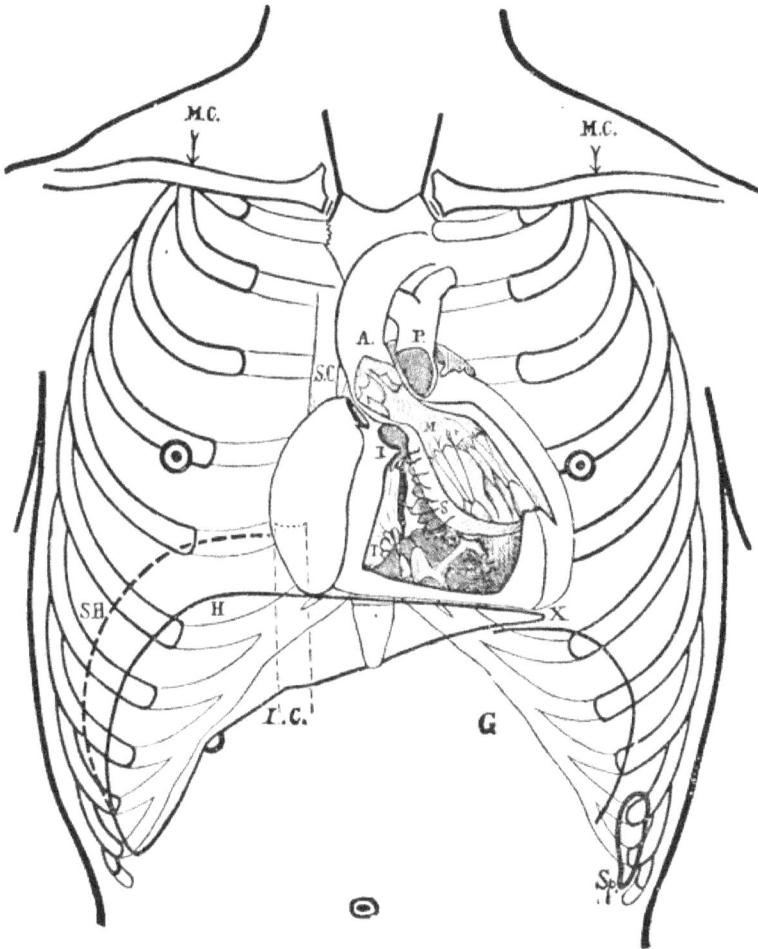

S—The foot of the septum, to the right and left of which are seen the ventricles. M—The anterior mitral flap. T—The right lower tricuspid flap. I—The conus arteriosus divided and partly removed. P—The pulmonary artery; behind P the left auricle is seen in outline. A—The aorta cut open, showing the right posterior aortic flap and part of the other flaps.

33

(10) The Mitral valve belongs to a yet more remote plane, being indeed nearer to the spine than to the sternum.

(11) As a whole the left ventricle occupies a higher level in the thorax than the right ventricle ; and

(12) The Mitral valve is situated high up, as well as far back in the ventricle.

THE LEFT AURICLE.

The Left Auricle is strictly posterior and entirely out of sight in a front view of the heart, with the exception of the tip of the auricular appendix which comes forward to the left of the root of the pulmonary artery.

In order to facilitate for the student an appreciation of its situation, the auricle is brought " diagrammatically " into view (" anatomically " it would be invisible) between the pulmonary artery and the aorta, and between the aorta and the superior vena cava.

Unlike the Right Auricle it does not rest on the diaphragm, but extends rather higher than its fellow,—in the same way as the left ventricle is placed at a higher level than the right.

It is the hindermost of all the cavities of the heart, and nearly fills the angle formed by the two main bronchi.

The Pulmonary Veins are entirely out of sight.

The position of the left auricular appendix affords demonstration of the left-sided position of the mitral orifice ; for in any heart the appendix lies above, and in close proximity to this orifice, and may therefore be taken as a guide to its situation.

THE POSITION OF THE GREAT VESSELS IN CONNECTION WITH THE HEART.

In addition to structures previously described, this Outline displays the origin and early course of the great vessels issuing from the heart, with the exception of the pulmonary veins.

THE AORTA.

(1) The Aorta is entirely hidden at its origin by the anteriorly placed Conus Arteriosus and Pulmonary Artery, and slightly by the Right Auricle.

(2) Within the pericardium it is exposed to view in its *Ascending portion* which is in right lateral and in anterior contact with the Superior Vena Cava.

(3) Behind the right second interspace, whilst still within the pericardium, it becomes yet more superficial, being only covered by a thin edge of lung, and its curvature is already pronounced. The part projecting beyond the right border of the sternum is, however, very small.

(4) Still smaller is the portion covered by the right 2d cartilage.

(5) Above this level the ascending portion of the arch lies entirely behind the manubrium, giving off the Innominate Artery on the right, and the Left Carotid and Subclavian on the left. At their origin these vessels are in close relation respectively with the right and with the left 1st chondrosternal junctions.

(6) The termination of the *Transverse portion* and the beginning of the *Descending portion* are seen in foreshortening in the left second interspace.

(7) The part of the Aorta just described lies almost entirely behind the upper sternum.

(8) It will be noticed that, whilst the first part of the vessel moves forwards and to the right, as far as the right 2d space, the second part recedes from the surface upwards and to the left.

(9) The highest level reached by the Arch is usually that of the lower border of the 1st cartilage. But there is much individual variety in this respect.

THE SUPERIOR VENA CAVA.

(1) The course of this vessel, SC, is for the greater part intrapericardial, usually as far as a point just below its bifurcation into the Innominate Veins.

(2) Its orifice into the right auricle, not shewn in the diagram, is nearly in the same vertical line as that of the Inferior Vena Cava (from which it is separated by the Eustachian Valve). It is deeply situated, but not so deeply as the orifice of the Inferior Vena Cava.

(3) Emerging from behind the base of the auricular appendix, the first portion of the Superior Vena Cava is somewhat overlapped by the ascending part of the Aorta in front.

(4) But, at the level of the 2d rib, whilst the Artery begins strongly to recede, the Vein advances, and, casting off its pericardial covering, divides into the vertical Right Innominate Vein and the horizontal Left Innominate Vein. The subsequent bifurcation of each of these two veins occurs behind the corresponding sternoclavicular joints.

THE INFERIOR VENA CAVA.

(1) The orifice of the Inferior Vena Cava, IC, into the right auricle is shewn in dotted outline. It lies far back, behind the thickness of the liver, and is raised, as it were, by the vertical thickness of the organ. It is, as stated above, the highest spot in that part of the diaphragmatic surface which is included within the pericardium ; and it lies higher therefore than any other spot belonging to the inferior surface of the right auricle or ventricle.

(2) The level of the orifice, in the parasternal line, corresponds to that of the 5th cartilage.

(3) The portion of the Vena Cava which projects above the diaphragm, purposely depicted here as a short vessel, is hardly recognisable as an independent structure, but immediately broadens out into the auricle.

(4) The posterior wall of the vessel is, however, more readily recognised than the anterior, as distinct from the auricle.

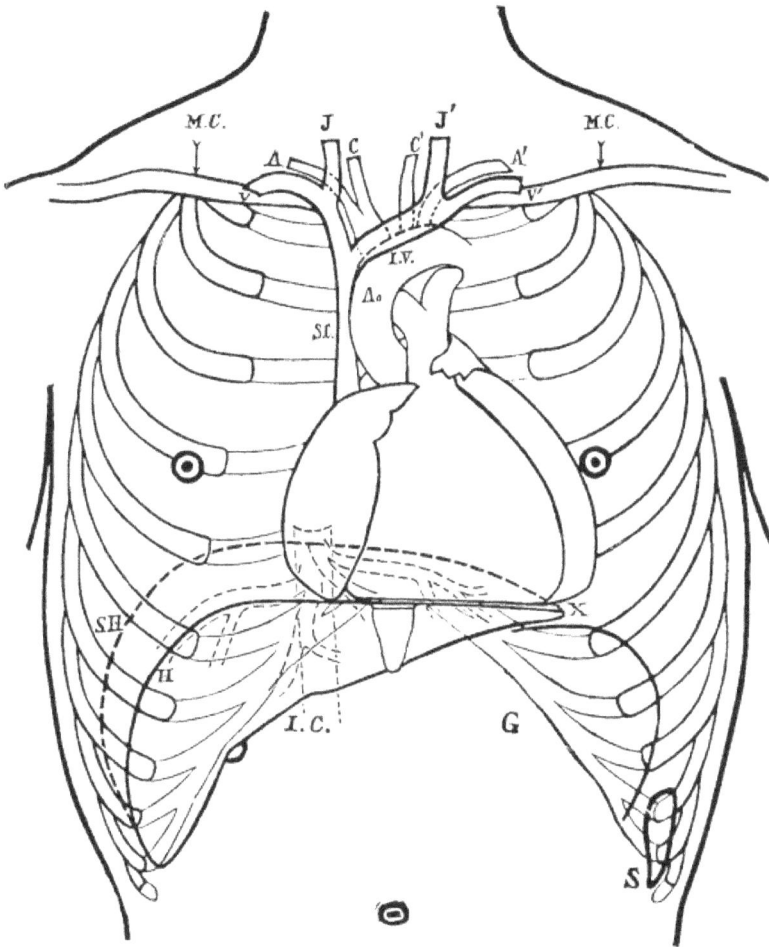

SC—The superior vena cava. Ao—The aorta. IV—The left innominate vein lying in front of the origin, from the arch, of the three great
vessels. JJ'—The jugular veins. AA'—The subclavian arteries. VV'—The
subclavian veins. CC'—The carotids. IC—The inferior vena cava receiving the hepatic veins.

THE PULMONARY ARTERY.

This vessel is remarkable for :

(1) Its shortness,

(2) Its almost vertical direction,

(3) Its superficial position,

(4) Its contact with both Auricular Appendices,

(5) Its funnel-shaped Conus Arteriosus.

Moreover for :

(6) The angle formed by its axis with that of the Aorta behind ;

(7) Its strongly lateral position—(it lies for the greater extent outside the left border of the sternum in the 2d interspace) ; and

(8) Its bifurcation behind the left 2d chondrosternal joint.

(9) The Right Pulmonary Artery passes nearly horizontally to the right, beneath the Arch of the Aorta.

(10) The Left Pulmonary Artery continues its upward course, with slight obliquity towards the left, and with increasing obliquity backwards. It finally arches downwards into the left lung, under cover, from above, of the Aortic Arch.

PART II.

THE DERMOGRAPHIC METHOD AND THE CONSTRUCTIVE SERIES OF OUTLINES.

The following Outlines are termed " constructive " because intended to shew how the normal cardiac tracing may be built up on data derived from the thoracic skeleton, or from superficial landmarks, with the help of the dermographic method, or method of surface markings.

The first tracings to be practised by the beginner are the *thoracic lines of reference*. Each student should make out by inspection and palpation and draw on the chest of a suitable subject the mid-clavicular, the parasternal, and the hepatic lines. Let this be done at least once thoroughly.

Tracings of the cardiac boundaries may be obtained in two ways :

(1) By the use of definite landmarks based on a knowledge of the relations of the heart to the thoracic parietes ; and

(2) More directly by means of percussion in each individual case.

The latter method, the only one available in pathological conditions, is essentially a clinical method, implying considerable skill and experience. Special attention will be given to it in Part IV.

On the other hand, the constructive method is applicable only to the normal subject ; and its uses are chiefly educational. It is a means of impressing the cardiac boundaries on the memory, and a useful introduction to the more difficult stages in cardiac study.

This part of the subject has been treated more fully than is necessary in the case of the junior student ; but the reader will easily select among the Outlines those best suited to his own requirements.

39

In the construction of the Cardiac Outlines the following sequence has been adopted :

(1) Determination of the position of the apex of the left ventricle.
(2) Determination of the base of the left ventricle.
(3) Determination of the left lateral cardiac boundary.
(4) Determination of the right lateral cardiac boundary.
(5) Localisation of the great vessels.

The *landmarks* used for this purpose are, as already stated, either superficial (cutaneous) or deep (thoracic or costal). Two constructive methods are thus at our disposal : the " superficial," more rapid but rather more sketchy method ; and the " deep," more laborious, but probably more exact one. Both will be described in connection with the Outlines.

HOW TO DETERMINE THE APEX SPOT AND THE LEFT VENTRICULAR BASIC LINE, OR LINE OF THE APPENDICES.

(1) Find the base of the Xiphoid, or the Infrasternal angle (formed by the 7th cartilages, which nearly meet at the extremity of the gladiolus).
(2) Through this draw a horizontal line H H' (the Hepatic line) ;
(3) And through the right mamilla draw the nipple line, which will meet the hepatic line at H.
(4) As a tangent to the areola of the nipple draw a vertical line, the Inner Areolar line. This will meet the hepatic line at Ar.
The *apex of the ventricle* will very closely correspond to the apex of the right angle thus formed.
(5) Draw the horizontal line MS, half-way between the episternal notch and the infrasternal angle, and
(6) Join the point H with the middle point of the sternum.
(7) The prolongation SL, of this line to the left is the line wanted, viz., the left ventricular basic line, or line of the appendices.

FIG. XIV.

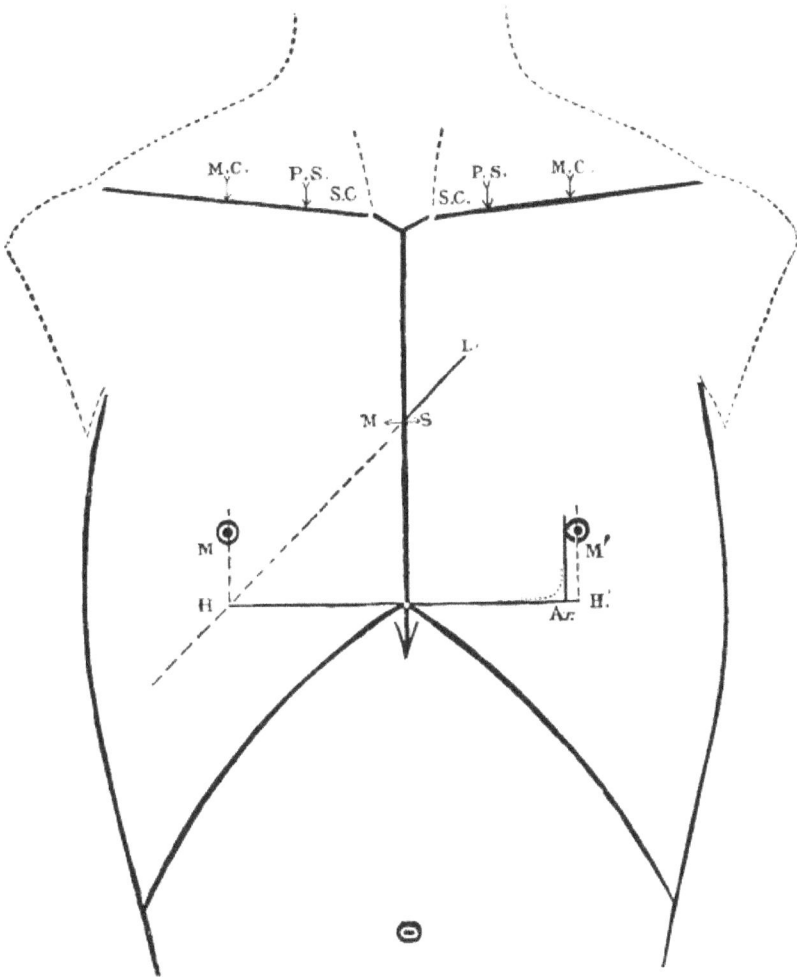

SC, SC'—Sterno-clavicular joints.   Ar—Vertical inner areolar line (the position of the cardiac apex is indicated in dotted line).   MM'—Nipples and nipple lines.   HH'—Hepatic line.   IIL.—Line conducted through the middle point of the sternum.   MS—Horizontal line dividing the sternum into two equal parts.

OTHER METHOD FOR DETERMINING THE APEX SPOT AND THE LEFT VENTRICULAR BASIC LINE, OR LINE OF THE APPENDICES.

This method differs from the preceding in that it relies upon the skeletal instead of the superficial landmarks. It requires as a preliminary that the ribs be recognised and a few of them numbered for reference.

Draw, as in the previous Outline :

(1) The hepatic line.

(2) The left inner areolar line (this line is seen to coincide exactly with the 5th chondrocostal junction—a valuable skeletal landmark for localising the cardiac apex).

(3) Find the right 5th and the left 2d chondrocostal junction :

(4) A line drawn through these two points will meet the middle line at MS, halfway down the sternum ; its prolongation to the left is the line sought.

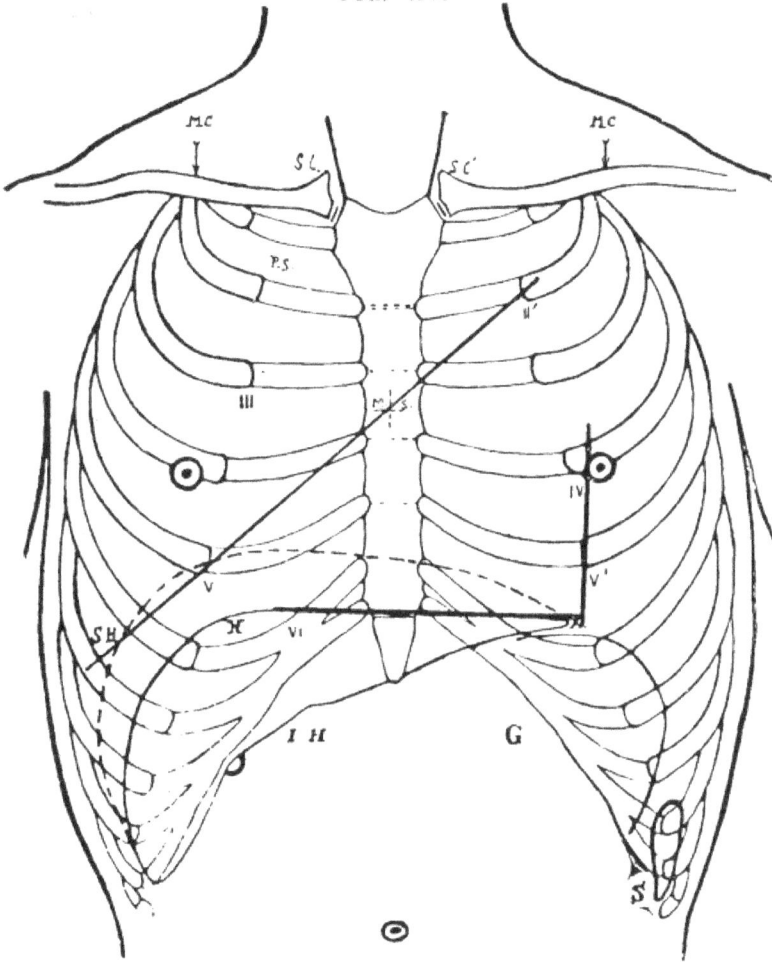

FIG. XV.

III—Right third costo-chondral junction. V—Right fifth costo-chondral junction. VI—Right sixth cartilage. MS—Line dividing the sternum into an upper and a lower half. II'—Left second costo-chondral junction. IV'—Left fourth costo-chondral junction. V'—Left fifth costo-chondral junction.

THE LEFT VENTRICULAR BASIC LINE, OR LINE OF THE APPENDICES, AND THE SITE OF THE PULMONARY ARTERY.

If a line M′SC, to which we may refer as the *left oblique line*, be conducted from the inner side of the left areola, to the right sternoclavicular joint, this will intersect at L the line II I, which was described in Outline XIV.

Divide the distance ML into three equal parts ; and draw at the points of division two lines of the shape and length of the small dotted lines. These will represent, as is shown in the following Outline, the sides of the Pulmonary Artery.

The further use of the line M′SC will be given on page 48.

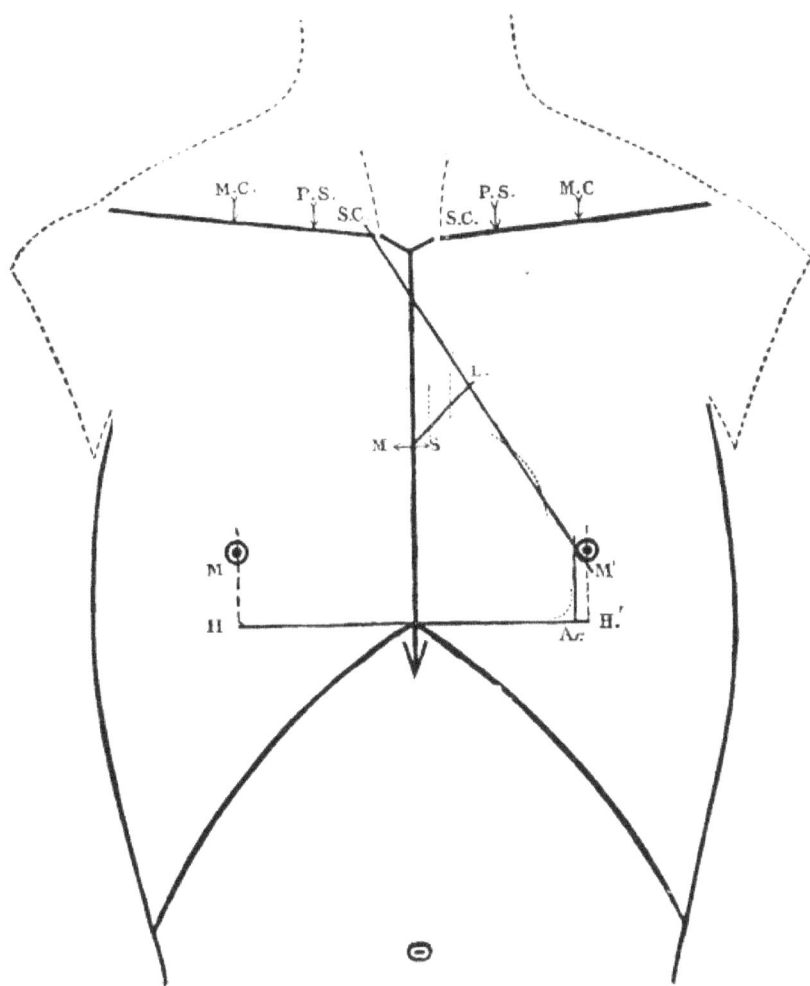

M'SC—The left oblique lateral line (the outline of the left ventricle is indicated in dotted line). ML—The line of the auricular appendices (cf. Fig. XVII.). The small dotted lines shew the position of the pulmonary artery. The other letters as in previous outlines.

45

HOW TO LOCALISE THE AURICULAR APPENDICES, THE PULMONARY ARTERY, THE AORTA, AND THE SEMILUNAR VALVES.

The line ML enables us to localise six important structures :

(A) THE TWO AURICULAR APPENDICES.

The Outline of the Pulmonary Artery is supposed to have been drawn as
described over leaf ;

At its sides, and slightly overlapping it, should be drawn the apex of the
Right and that of the Left Auricular Appendix. The two appendices exactly face each other, and they both terminate (AS, AP)
on the line ML.

Each auricular appendix occupies a lateral third of the line ML. In
shape they may be roughly depicted as triangles, the base of which
would be perpendicular to the line ML.

Further peculiarities of outline should be committed to memory.

It is understood that this description is purely diagrammatic. The tips
of the auricular appendices (especially that of the left), may deviate from
the line ML, and may differ slightly in length and in direction from those
here depicted.

(B) THE AORTA AND THE AORTIC SEMILUNAR VALVE.

The axis of the first part of the Aorta is very nearly perpendicular to the
direction ML., and the vessel rises, as it were, partly from the middle third,
and partly from the inner third of that line.

The aortic orifice and semilunar valve thus coincide, in their transverse
axis, with the line ML.

The anterior aortic semilunar flap AS is dotted to signify that it is situated behind the pulmonary artery.

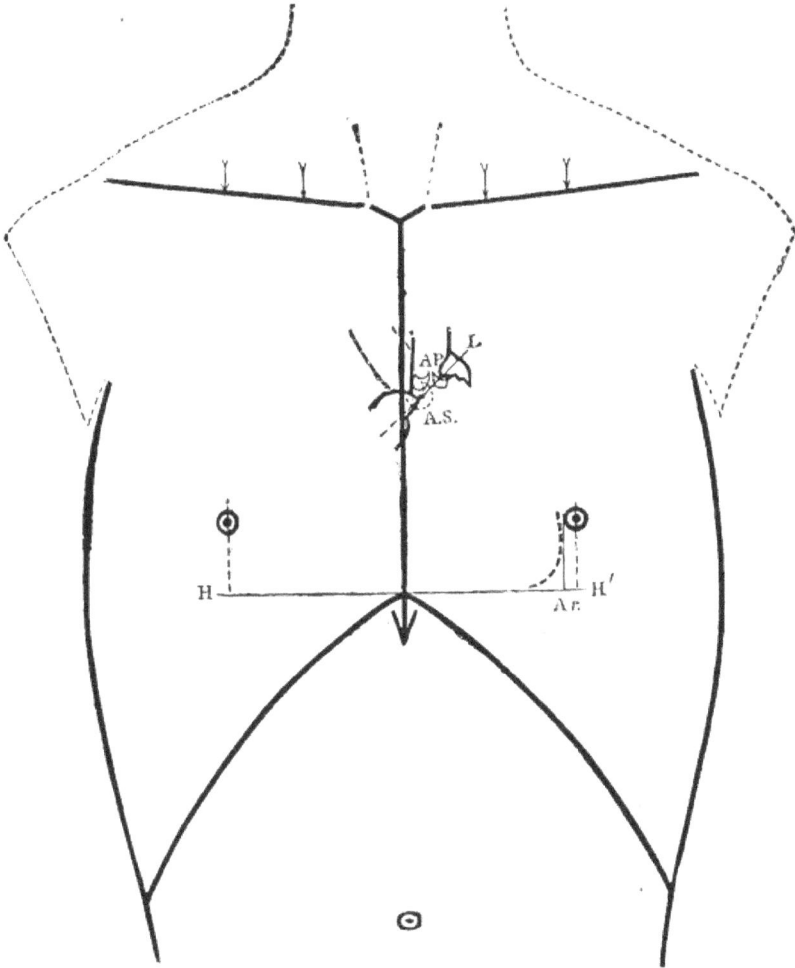

FIG. XVII.

HH'—Hepatic line. Ar—Vertical inner areolar line. L—Left auricular appendix, facing which is the right auricular appendix. AP'—The two anterior pulmonary valve flaps. AS—The anterior semilunar flap of the aortic valve, shewn in dotted line to indicate its posterior position.

47

(C) THE PULMONARY ARTERY AND THE PULMONARY SEMILUNAR VALVE.

The Pulmonary Artery is vertical in its first portion.

Its orifice and its valve are therefore horizontal ; and their level is above the main level of the aortic valve. The latter, however, at its left extremity, just rises to the level of the pulmonary valve. In the diagram are depicted at AP the right and the left anterior pulmonary valve flaps.

———

FIRST METHOD OF TRACING THE LATERAL LINES OF THE CARDIAC DIAGRAM.

(1) Both left lateral lines, Ar and M'Sc, have already been described.

(2) The right lateral lines are also two :

　　(a) The parasternal line, in its lower third ; and

　　(b) The vertical line C, which may be termed the line of the Vena Cava ; this line is parallel and internal, by about ½ inch, to the parasternal line.

(3) The line AOSC is drawn parallel to HL, and passes through the left sterno-clavicular joint ; it roughly defines the upper limit reached by the ascending Aorta.

(4) Lastly, the line Au Au', also parallel to HL and distant from it only one inch, is the least essential of all ; it gives the right upper boundary of the right auricle, and of the heart.

The slight difference observed between the shape of the heart in this Outline and the most of the others is due to the difference in the shape of the thorax, which in them has purposely been made very broad, and also to the somewhat low position given to the nipples in Outline XVIII. The correctness of the directions is not affected by the imperfections of the drawing.

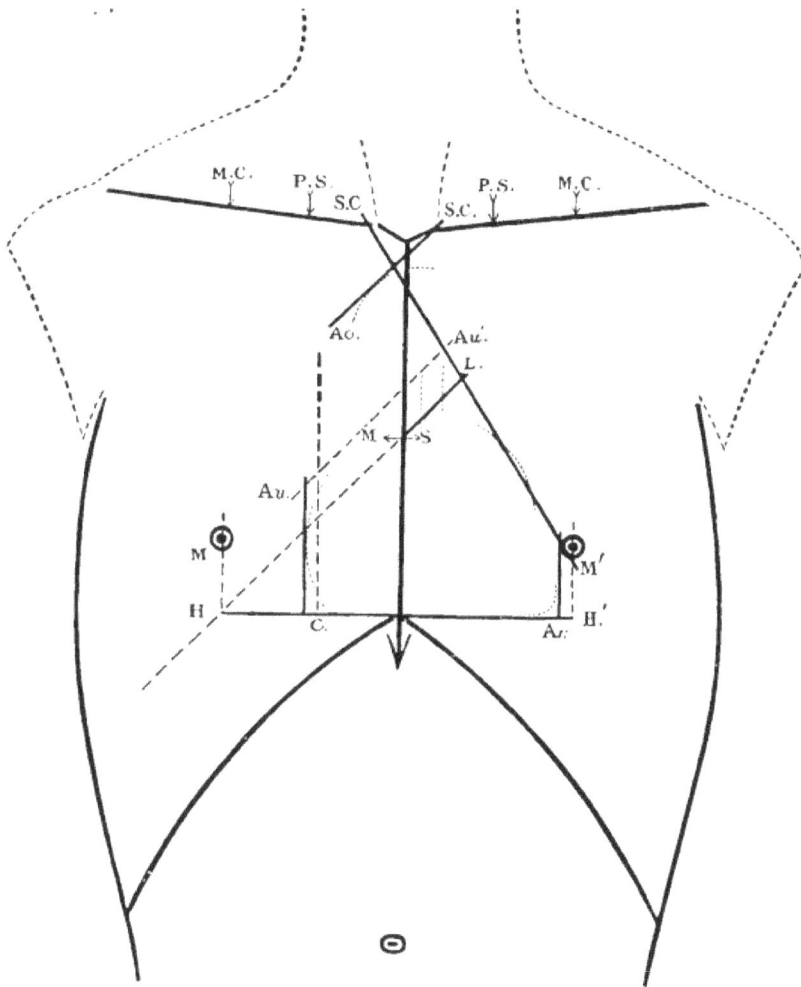

FIG. XVIII.

Ao—Outline of the arch of the aorta. AoSC—Right upper boundary of the arch. Au.Au'—Right upper boundary of the heart. C—Line of the vena cava, external to which is the right lateral boundary of the heart, nearly coinciding with the parasternal line, and the dotted outline of the right auricle.

49

SECOND METHOD OF TRACING THE LATERAL LINES OF THE CARDIAC DIAGRAM.

The lateral lines are in this case determined with the help of skeletal landmarks.

(1) A vertical line drawn through the 5th chondrocostal junction is identical with the line described as the vertical inner areolar line ; and

(2) A line drawn through the right sterno-clavicular joint, and through the left 4th chondrocostal junction, is the same as the oblique line obtained with the areola as a landmark.

(3) The parasternal line (dotted) needs no comment.

(4) The line of the Vena Cava may be drawn from the middle of the right 2d cartilage, downwards ; or else, upwards, from the junction of the hepatic line with the upper border of the 6th cartilage, or more simply according to the rule given under Outline XVIII.

(5) The upper line is obtained by joining the right 3d chondrocostal junction with the left sternoclavicular joint.

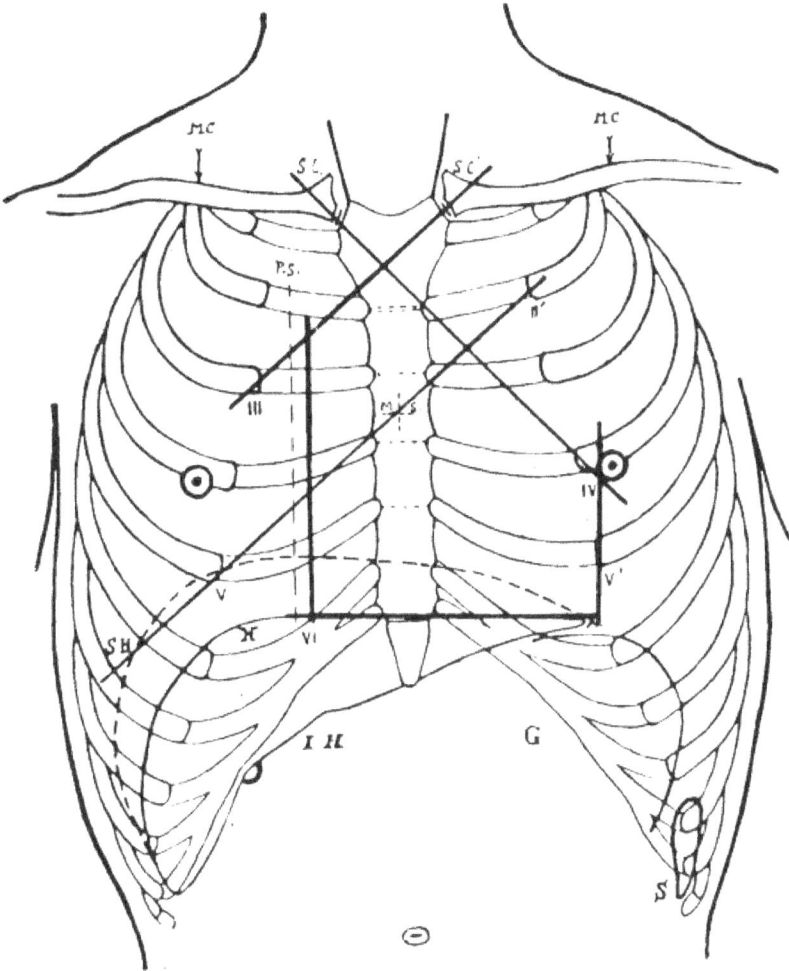

FIG. XIX.

The same meaning attaches to the letters and numerals as in Fig. XV. IIISC—Right upper boundary of the aortic arch. IVSC—Left oblique line. VI—Line of vena cava. PS—Parasternal line, nearly coinciding with the right lateral cardiac boundary.

THE HEART'S OUTLINE COMPLETED ACCORDING TO THE FIRST METHOD.

(1) The right lateral boundary, belonging exclusively to the right auricle, is continued downwards from the upper part of the right auricular appendix in a broad curve which touches the line Au and the parasternal line.

(2) Line C gives the outer outline of the Inferior Vena Cava; and the auricular orifice of the vein is indicated by the 6th cartilage.

(3) The left lateral cardiac outline forms a yet broader curve than the right. This curve extends a little beyond the oblique line, but barely touches the vertical inner areolar line.

(4) The Aortic Arch rises into contact with the line AOSC described in Outline XVIII.

(5) Under the Arch passes, with almost horizontal direction, the right branch of the pulmonary artery, the left division of which, at first almost vertical, ultimately bends over the left bronchus and under the Arch.

(6) The right auriculo-ventricular boundary passes obliquely downwards from the middle point of the sternum to its outer edge close to the xiphoid.

(7) The interventricular line extends downwards with gentle curve from the left auricular appendix to the cardiac apex.

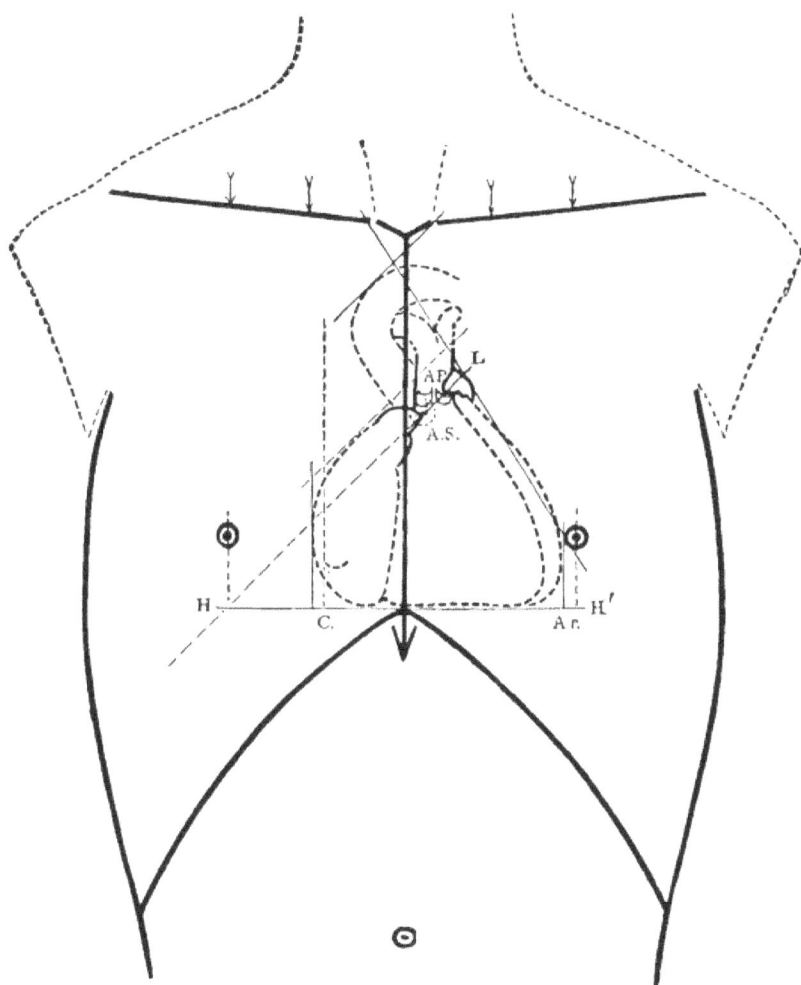

The letters have the same meaning as in previous Outlines. Compare Outlines XIV., XVI., XVII., and XVIII.

THE HEART'S OUTLINE COMPLETED ACCORDING TO THE SECOND METHOD.

The slight difference in shape noticeable between this cardiac Outline and the preceding one, Fig. XX., mainly arises from the differences which may be traced between the relative length and width of the two thoracic diagrams, and between the relative aperture of the infrasternal angles. In reality the shape of the heart does not vary much ; but its position and that of the great vessels, and the shape of the "presenting" cardiac surface, are subject to variations to suit thoracic space. A comparison of the two Outlines will remind the student of the importance which attaches (from the point of view of cardiac study) to a careful observation of the individual peculiarities in the build of the thorax.

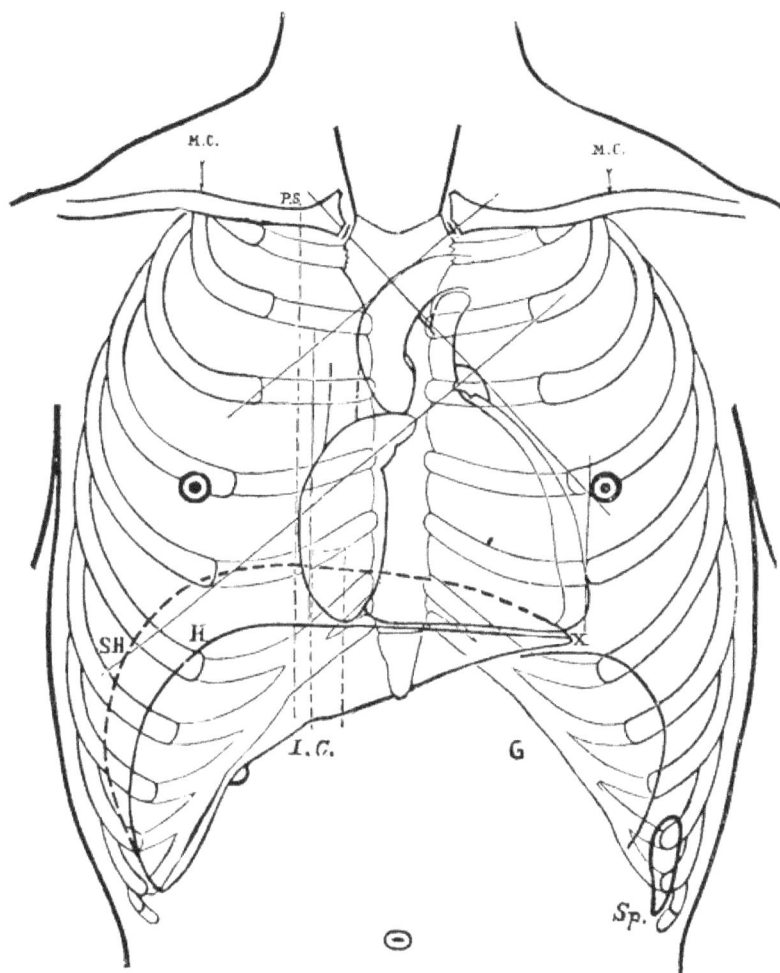

The letters have the same meaning as in previous Outlines. Compare
Outlines XV. and XIX.

PART III.

THE PRACTICAL METHODS OF INSPECTION AND PALPATION.

This is a short account of the subject, intended for the beginner. It is not illustrated with any diagrams or outlines.

I. INSPECTION OF THE ANTERIOR SURFACE OF THE THORAX AND OF THE PRÆCORDIUM.

Observe any peculiarity of the anterior thoracic surface :

(1) *Any irregularity in outline*
  (a) of the clavicles,
  (b) of the sternum,
  (c) of the costal arch and infrasternal angle,
  (d) of the xiphoid appendix ;
(2) *Any bulging of the præcordium ;*
(3) *Any asymmetry of the nipples—*
  (a) in the vertical direction,
  (b) in the horizontal direction.

   One or the other form of asymmetry of the nipples is often associated with uneven bulging of the præcordium.

(4) Notice carefully and mark on the chest *the sites of any visible pulsation* that may be present in the episternal, in the infrasternal, or in the intercostal spaces ; and especially *the site of the apex beat.*

56

II. PALPATION OF THE ANTERIOR SURFACE OF THE THORAX AND OF THE PRÆCORDIUM.

THE COSTAL CARTILAGES AND INTERSPACES.

Our first object in feeling the chest is to ascertain the position of the several ribs and interspaces. The experienced clinical worker can place his finger at once upon any rib or interspace he may be desired to find. The beginner has to search for them. With the help of the following methods this can be done successfully and quickly.

Counting the Costal Cartilages.—

(1) Count from above downwards, and in the vertical nipple line.

(2) For safety sink the edge of the thumb into the interspace below the cartilage, whilst the index finger is still on the latter, so as to always keep touch with a cartilage and with an interspace.

(3) Whilst the thumb passes from the interspace to the next cartilage below, let the index take its place in the interspace.

(4) Alternate, after this fashion, the position of the index and of the thumb until the costal arch is reached.

(5) The accuracy of the count is confirmed when the last costal cartilage, ending in contact with the xiphoid cartilage, is found to be the 7th.

How to find the First Costal Cartilage and Interspace.—

(1) Feel with the fingers for the sternal end of the clavicle, which it is impossible to mistake.

(2) Pointing the index, endeavour to push it upwards beneath the head of the clavicle ; it will hitch against the 1st costal cartilage, which may then be followed for some distance outwards until it recedes beneath and behind the middle third of the clavicle.

(3) Meanwhile the pulp of the finger will recognise, beneath the cartilage, the soft resistance of the 1st interspace.

(4) It is a distinctive feature of the 1st costal cartilage that the finger cannot reach its upper border.

How to find the Second Cartilage.—By some the following method is adopted in preference to the preceding one :

(1) Draw the flat of the finger vertically along the manubrium, in the middle line, till it encounters the *transverse ridge* formed by the cartilaginous union of that bone with the gladiolus. The 2d cartilage is situated in a line with the ridge, articulating with both bones.

In thin subjects the sternal end of the second cartilage is generally prominent and conspicuous ; and the transverse ridge mentioned above may often be seen independently of palpation.

In order to make quite sure of the 2d cartilage it is well to press the finger into the interspace above it, and to feel the 1st cartilage in close relation with the clavicle.

Another simple method, applicable to thin subjects, is to run the flat of the index finger down the chest in the nipple line, so as to distinctly feel the cartilages. The highest cartilage of which the upper border is thus felt, is the 2d. The 1st cartilage, as previously stated, only admits of its lower border being felt.

The Seventh Cartilage and the Costal Arch.—The remaining cartilages present no difficulty until the 6th is reached. This cartilage and the 7th are a little difficult to distinguish from each other in the neighbourhood of the sternum. The 7th may, however, be easily identified from below, inasmuch as it is

(1) the last cartilage that can be felt by the finger at the lower part of the sternum and

(2) the cartilage which forms, with the xiphoid, the infrasternal angle.

(3) If, therefore, the finger be placed on the xiphoid, the cartilage felt on either side of it is the 7th.

Explore the infrasternal angle and the sides of the costal arch, noticing the 8th, 9th, and 10th junctions, and below them the tip of the 11th rib.

Observe the height and the width (or aperture) of the arch ; whether or not symmetrical, etc. Ascertain the shape and the size of the xiphoid appendix, and any deformity or malposition it may present.

THE NIPPLE.

Lastly, find the thoracic site of the nipple. This will be in front of, or immediately below, the 4th rib, about an inch external to its junction with the cartilage. Notice whether or not the mid-clavicular line passes through the nipple.

III. PALPATION OF THE PRÆCORDIUM FOR CARDIAC IMPULSES.

The præcordium should once more be explored by palpation as to the existence of any bulging (frequently associated with asymmetry of the nipple) and of any pulsation or thrill.

Pulsation should be felt for

(a) at the apex,
(b) at the infrasternal angle,
(c) at the episternal notch, by hooking the index and median fingers gently into it ; and
(d) over the præcordium in general, by laying the flat of the hand over the front of the chest.

THE APEX-BEAT.

This may be quite palpable, or may convey to the finger only a distant pulsation. The beat may happen to be situated behind the rib. In describing it notice should be taken of

(1) its exact position,
(2) whether it be strong or weak,
(3) small or extensive,
(4) prominent or retractile.

PULSATIONS OTHER THAN THE APEX-BEAT.

Pulsation at the epigastrium is common. It is frequently transmitted ; sometimes it is direct.

Pulsation is often observed in the left 4th interspace. In this and in the following situation it is generally spoken of as diffused by contrast with the more circumscribed beat at the apex.

Pulsation in the 3d left interspace often accompanies that in the 4th.

Pulsation may also occur in the 2d left interspace. It may be *systolic* in time.

There is however another form of pulsation, sometimes seen at the left 2d interspace, namely, a *limited and diastolic* impulse. This is due to the diastolic tension of the sinuses of Valsalva of the pulmonary artery.

Pulsation does not occur on the right side of the chest, except when the heart is displaced or transposed, and in cases of aneurysm.

Pulsation in the episternal notch may be felt at varying depths ; sometimes it may be felt or even seen at the upper level of the manubrium.

IV. PALPATION IN PATHOLOGICAL CONDITIONS.

ABNORMAL IMPULSES AND THRILLS.

In abnormal subjects other impulses may be felt besides that of the cardiac apex ; and the latter may itself be abnormal.

ABNORMAL APEX-BEAT.

The chief abnormalities of the apex-beat relate to its position, to its strength, and to its extent.   Thus the apex-beat may be

Displaced, or

Too strong, or

Too widely felt (extensive apex-beat).

The heart's apex is so frequently hidden *behind the 5th or the 6th rib* that the mere fact that it cannot be felt beating does not warrant a conclusion that any abnormality exists.

ABNORMAL CARDIAC IMPULSES.

Cardiac impulses may be :

(1) Direct ;

(2) Indirect or transmitted.

One form of *transmitted* impulse is of frequent occurrence, viz., the *indirect epigastric* pulsation conveyed by the liver.

Most præcordial impulses are *direct*.   Moreover, there frequently occurs a *direct epigastric* cardiac impulse which it is important not to overlook.

The præcordial impulses are systolic or diastolic in time, and auricular, ventricular, or vascular in their origin.

The following table gives a useful synopsis of the impulses which may be felt over the heart or in its vicinity :

Palpable or Tactile Impulses.	Indirect or transmitted.	Transmitted epigastric pulsation ; Pulsations conveyed by tumours or fluid effusions ;
	Direct.	Apex-beat (prominent or retractile) ; Left Ventricular impulse ; Right Ventricular impulse ; Right Auricular impulse ; Epigastric (infrasternal) impulse ; Pulmonary systolic impulse ; Pulmonary diastolic impulse ; Aortic systolic impulse ; Aortic diastolic impulse.

A discussion of the mode of production and of the diagnostic value of the various cardiac impulses cannot be attempted in these pages. The same remark applies to the thrills, which may be classified as follows :

ABNORMAL PRÆCORDIAL THRILLS.

Thrills may be produced within the heart or outside it ; in other words, they may be *endocardial* or *exocardial*.

ENDOCARDIAL THRILLS may be *localised* to one part, or *general*—that is, felt over the whole præcordium.

Non-localised or general endocardial thrills are usually either congenital, or due to aortic valvular stenosis, or to aortic aneurysm ; and in all these cases they are usually systolic.

Localised endocardial thrills may be due to defect of any of the heart valves, although uncommon in connection with the right heart. A thrill may occur also when the valves are normal, as, for instance, in anæmia or as a result of pressure.

A *mitral* valvular thrill is usually diastolic or præsystolic ; an *aortic* valvular thrill more commonly systolic.

EXOCARDIAL THRILL is usually the result of *pericardial* friction. It may occasionally be produced by friction within the præcordial portion of the *pleural* sac (pleuro-pericardial friction).

PALPATION FOR IMPULSES AND THRILLS IN PERICARDITIS.

In view of its importance this subject claims to be separately considered :

IN FIBRINOUS PERICARDITIS a *friction-thrill* is usually perceptible, and is often very marked. It may be *general,* or *limited* to a portion of the præcordium. When of limited extent, it is more commonly felt over the middle than over the lower third of the sternum. When faint, it may be intensified by pressure. Pericardial friction-thrill is not conducted beyond the seat of its production. This is a useful point of distinction between it and some of the vascular thrills.

IN ADHESIVE PERICARDITIS palpation does not afford us much assistance. *Systolic retraction* of the region of the apex is not, in itself, a reliable sign. When associated with diastolic impulse it acquires greater diagnostic value. The late secondary changes undergone by the heart are to a certain extent perceptible by palpation, but they are not distinctive of the condition.

IN PERICARDITIS WITH EFFUSION there may be noticed on palpation, if the collection of fluid be a large one, *some bulging* of the whole dull area and of the intercostal spaces within it. *The cardiac impulse may be totally absent,* or perceptible only when the patient lies down ; *or it may be readily felt. A cardiac impulse occurring above the lower limit of the dull area* is diagnostic, especially if change in position should modify the shape of the dull area according to the laws of gravitation.

PART IV.

CARDIAC PERCUSSION AND THE "PERCUSSION"
SERIES OF OUTLINES (NORMAL AND
PATHOLOGICAL).

The importance and difficulty of this section of cardiac study have necessi-
tated the introduction of a set of graduated Outlines, illustrating step by
step the method to be followed, and the results to be obtained.

GENERAL REMARKS ON CARDIAC PERCUSSION.

The reader is supposed to be familiar with the general principles and
methods of percussion, and these apply to the physical examination of the
heart as they do to that of any other organ.

A great deal can be accomplished by unaided *finger percussion ;* and
until this mode of percussion has been mastered, the student should attempt
no other. For the purposes of the more advanced clinical worker *Sansom's
pleximeter* possesses decided advantages, which will be explained presently.

The special need in cardiac percussion is *precision.* We should be able
to obtain an accurate idea of the size of the heart, in all its diameters,
except the antero-posterior ones. This can be done, in spite of the constant
alterations in size and in shape which the organ undergoes, since, owing to
the short duration of the systole, we are for practical purposes always percussing the heart during its diastole.

THE DERMOGRAPHIC METHOD.—Correct demarcation is the highest aim
of percussion ; it is also the quickest way to acquire precision. Endeavour,
therefore, to trace out the results of percussion on every suitable occasion.

Never be without a dermographic pencil, and begin to use it independently at your earliest clinical opportunity. Having made out your own markings, get one of your seniors to verify them. Then go over the work again, the better to perceive your mistakes and their corrections. Your progress may be rapid or it may be slow ; if it should be slow, so much the greater was your need for practice, and so much the greater will be your ultimate gain.

THE METHOD OF PERCUSSING.

ON THE USE OF THE SINGLE AND OF THE DOUBLE PERCUSSION STROKE. —In percussing the cardiac dulness in the dead body, a single stroke is enough to decide whether the spot percussed be resonant or dull. During life the movements of the lungs and of the heart are an excuse for repeating the stroke, since they might possibly interfere with a correct estimate of the amount of dulness or of resonance at any given spot.

Multiple percussion, which elsewhere is not to be recommended, is therefore advisable in cardiac percussion. By this, however, is not meant a confusingly rapid peal of taps, but two—or at most three—deliberate strokes, with time enough between them to appreciate the note struck at each blow.

SHORT RULES FOR SAFE AND RAPID PERCUSSION.

It is not enough to know how to map out the heart correctly ; we must learn to do it quickly, and yet reliably. The following methods will save time, and will be found useful even by advanced students ; whilst they are invaluable to the beginner :

(1) Before percussing the heart, *a standard of complete pulmonary resonance* must be sought at a safe distance from the præcordium, be it at the upper pectoral region, or in the anterior axillary line.

(2) *A standard of complete cardiac dulness* should next be obtained by percussing the 5th cartilage or interspace in the left parasternal line.

(3) For the rapid discovery of the line which separates complete pulmonary resonance from the partial cardiac dulness ("partial" owing to the influence of the thin covering of resonant lung), it is essential to bear in mind the shape of the cardiac outline which has been depicted in the preceding diagrams.

(4) The direction of the finger or pleximeter must in every case be parallel with that of the line for which we are, at the time, percussing. Therein lies the secret of accurate results.

(5) Thus, if we should want to define a *vertical* portion of the cardiac outline, the finger or pleximeter must be applied to the chest vertically ; if a horizontal portion of the outline be wanted, then the pleximeter must be horizontally placed. Let us imagine, for example, a case in which we are quite ignorant as to the distribution of the dulness and of the resonance, and in which their boundary is vertical—the ease with which this boundary will be found will vary with the direction given to the finger or pleximeter. If the flat of the finger be placed horizontally, any percussion of it will elicit a mixed sound arising partly from the resonant and partly from the dull surface ; however prolonged our percussion in this position, we shall be none the wiser. Should, however, the finger be applied vertically, the note obtained will then be that either of the dull or of the resonant area. By slightly shifting the finger (whilst preserving its longitudinal direction) we shall at once have an opportunity of appreciating the different notes special to the two surfaces.

A like success will attend our percussion of the several boundary lines of the heart's partial dulness, if only we are careful to percuss in directions parallel to each of them.

(6) The two alternatives would be : (a) either to percuss at each stroke a very limited area, no larger than the tip of the finger,—a time-consuming and perplexing task ; (b) or, if using the length of the finger, to endeavour by shifting its axis to find the position in which percussion yields a definite line of demarcation. We are spared all this trouble by adopting at once positions parallel to the lines of the normal cardiac outline.

N. B. Cammann's method of auscultatory percussion is not described in these pages, the ordinary method being simpler, and yielding results which are entirely satisfactory.

5

SANSOM'S PLEXIMETER.

Most pleximeters have this in common, that their chest-piece is straight, at least on one side ; and readily adapted to the direction of straight lines of dulness, such as have been described.

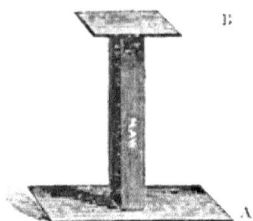

On the other hand, they all modify the sounds of percussion more than does mediate finger percussion.* This reproach also attaches to Sansom's pleximeter, although in a less degree than to any with which the author is acquainted.

Through Dr. Sansom's kindness he is able to give an illustration of the pleximeter. The instrument is made of vulcanite and measures 4 *cm.* in height ; its longer flange A 4 *cm.*, its smaller flange B 2 *cm.* only in length. The width of each flange is 13 *mm.*

FIG. XXII.

SANSOM'S PLEXIMETER.

WHAT TO AVOID IN CHOOSING A SANSOM'S PLEXIMETER.

The pattern seen in Fig. XXII is obsolete and should be avoided in making a selection. Its fault lies in the sharpness of the angles, which occasions pain to the patient.

The second illustration, Fig. XXIII, shews the corners rounded off. It also shews rather more substantial flanges than those seen in the original pattern. The purchaser should select flanges of medium thickness, just thick enough to resist the attempt to bend them. The very thin flanges bend readily, and are brittle ; neither are they so well adapted for their special purpose as rather thicker ones.

* For further remarks on this subject see the author's clinical lecture " On the value of accuracy in cardiac percussion," *Lancet*, August 29, 1891.

HOW TO USE A SANSOM'S PLEXIMETER.

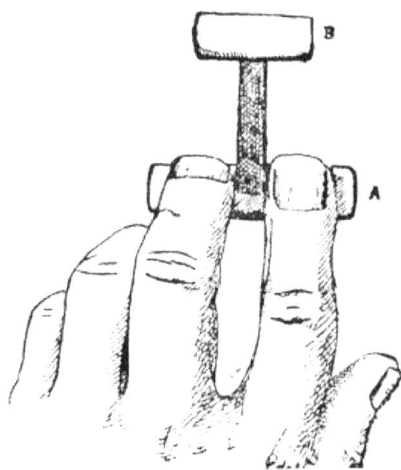

FIG. XXIII.
THE PLEXIMETER IN USE.

Little explanation is needed for Fig. XXIII., which shews the usual arrangement, viz., the long flange used as the chest-piece, the small flange as the anvil, and the finger-tip as the hammer.

The position can be reversed whenever a very small chest-piece is desirable. This facility is often a valuable help.

The figure represents the method (recommended by Dr. Sansom) of holding the instrument by the lateral pressure of the 2d and 3d fingers, their pulp resting on the flange, and being accessible through it to almost all the chest vibrations elicited by percussion.

ENUMERATION OF SOME OF THE ADVANTAGES OF THE PLEXIMETER.

Among the advantages special to Sansom's pleximeter may be mentioned :

(1) Its lightness,
(2) Its small size and portability,
(3) Its ability to fit in awkward hollows which some fingers can hardly reach,
(4) Its reversibility,
(5) The ease with which its pressure may be regulated by the finger,
(6) The straightness of its edges, which favours rapidity quite as much as might the same length of the flat of the finger, whilst affording superior sharpness in the results,

(7) The facility given for measurements, since millimetres or lines might be marked on the upper face of the flange.

(8) The inherent fault previously mentioned is—for cardiac percussion—also to be reckoned among the special advantages. This pleximeter intensifies dulness, at the same time that it defines more sharply the dull border.

Therefore, if doubt should occur as to the value of any sounds obtained by immediate finger percussion, the pleximeter would probably decide. In this power for discrimination lies a yet stronger recommendation than in the practical advantages which have been enumerated.

"PERCUSSION" SERIES OF OUTLINES.

THE CARDIAC SURFACE LEFT UNCOVERED BY THE LUNGS.

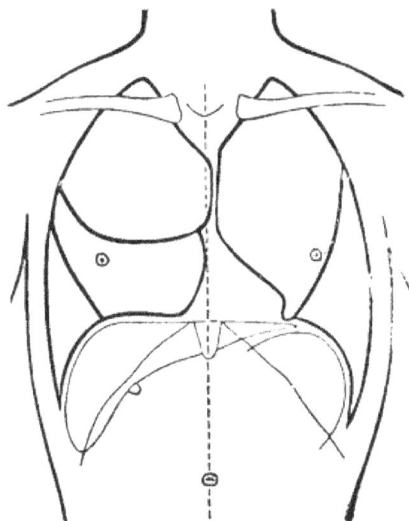

FIG. XXIV.

The accompanying illustration will remind the student of the relation existing between the anterior pulmonary borders and the heart, which is not depicted here, but which occupies the interval between them. The right and the left upper lobe come into close relation with each other behind the sternum on a level with the 2d and 3d cartilages. Below the latter they diverge. The fringe of the left upper lobe receding much farther and much more abruptly than that of the right, gives rise to the gap known as the cardiac incisure.

The portion of the præcordium devoid of any anterior pulmonary covering will be seen in Outline XXVIII. to correspond to the following parts :

The lower third of the sternum,
Part of the left 5th cartilage,
Part of the left 6th cartilage,
A small portion of the 4th interspace,
The greater part of the 5th interspace.

Over this extent of surface nothing but pericardium and areolar tissue intervenes between the heart and the anterior thoracic wall.

THE AREA OF ABSOLUTE CARDIAC DULNESS.

The heart, when isolated, being absolutely dull on percussion, absence of any pulmonary tissue over the space described in the previous diagram, would entail absence of resonance.   It is found, however, on percussing the chest, that absolute dulness *does not extend over the whole area* unoccupied by lung.   This is due to the facility with which the sternum conducts sound.   In its upper part this bone receives sonorous vibrations from the underlying lung tissue, and these are transmitted, almost unaltered, as far as its lower extremity.   Consequently the dulness obtained when the lower sternum is percussed is not, as it otherwise would be, absolute in degree. Absolute cardiac dulness begins only at the left border of the sternum ; this is its limit towards the right.   Its remaining boundaries are identical with those of the cardiac incisure of the left upper lobe.

ITS SHAPE AND SIZE.

These are apt to vary not only in disease, which we exclude from our present consideration, but also in the healthy chest, because the shape and size of the cardiac incisure are slightly variable.   On this point published descriptions do not all agree, some giving a circular, others a triangular shape as that of the normal area of absolute dulness.   This need not puzzle the student.   The question is one of anatomy, which he can settle for himself at any autopsy.   The shape of the cardiac incisure of the left lung is also the shape of the absolute dulness of the heart ; and in size they are also identical, with the exception mentioned above, that absolute dulness does not extend behind the sternum, although space is there occupied by the heart.   The boundaries of the area of absolute dulness are, therefore, in a majority of subjects :

laterally : (1) The left sternal border, and (2) A vertical line a little internal to the lower part of the left inner areolar line ;

above : (3) A line passing obliquely from the left 4th sternal junction to the lower border of the 5th cartilage, where the latter is crossed by the lateral boundary,

below : (4) The hepatic line.

The shape is seen to be a trapezoid or a truncated triangle.   For practical purposes the area of absolute dulness may be regarded as triangular.

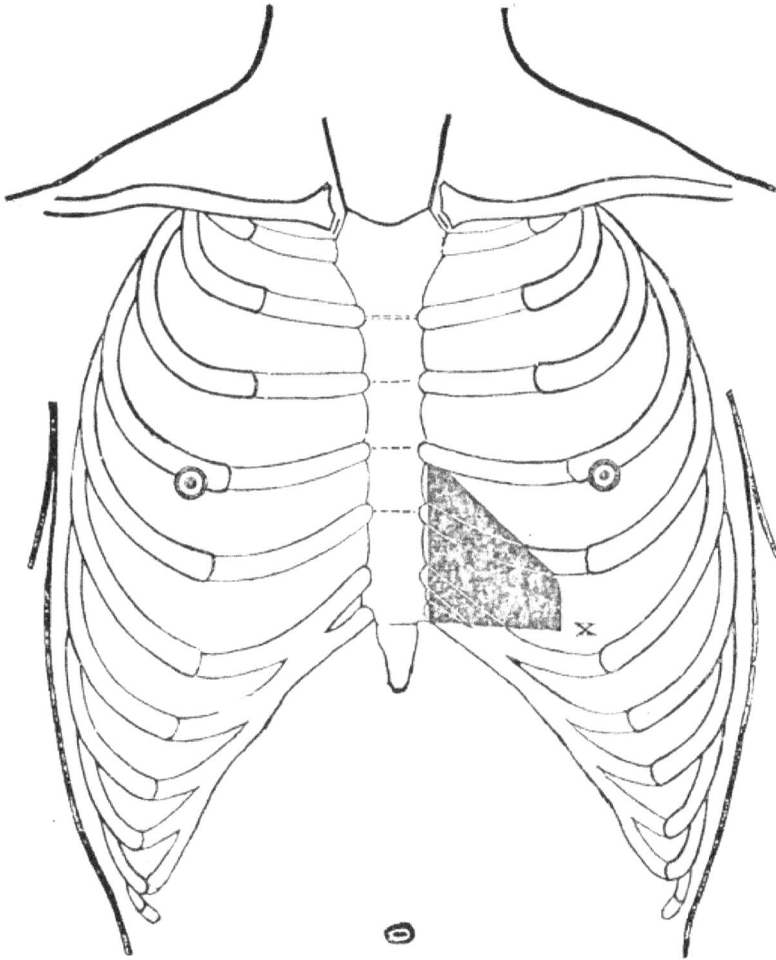

This Outline should be compared with Outline XXVI., and with Outlines showing the heart *in situ.* The left vertical boundary of dulness will be noticed to be slightly internal to the apex-beat X.

FIRST STEP IN THE EXAMINATION OF THE HEART BY PERCUSSION :

HOW TO DETERMINE THE ABSOLUTE HEPATIC DULNESS.

A preliminary percussion of the liver is the safest and, to the experienced clinical observer, also the shortest way of determining the area of cardiac dulness ; it has for its special object to find the line HX—the " hepatic line "—which corresponds to the level of the floor of the pericardium.

METHOD OF PERCUSSING.

(1) If possible, find the apex-beat, X, and mark the spot on the chest.

(2) Place the finger horizontally in the 4th right interspace, one inch internal to the right nipple line. This, on being percussed, will yield a fully resonant note.

(3) Transfer the finger to the 6th interspace, or 7th cartilage, in the same vertical line, still keeping it horizontal, and percuss again. This will give an absolute dull note, that of the liver.

(4) Percuss next in the 5th space, just below the 5th cartilage. The note will no longer be absolutely resonant ; but

(5) If the 6th cartilage, just internal to H, be now percussed, a much duller sound will result.

(6) To make out with precision the line HX is now merely a question of attentive percussion.

(7) Taking the highest point in the line HX, join it to a spot immediately beneath the apex-beat, X.

(8) By a similar process the line HH will be found.

(9) Any part of the outline included between these two lines, as far as the left border of the xiphoid cartilage, will yield an absolutely dull sound ; but

(10) *Beyond the xiphoid the hepatic dulness will cease to be absolute ;* and the degree of the change will vary according as the stomach, G, is inflated, full, or empty.

(11) The change is always from greater to less resonance as we pass from the right to the left portion of the liver.

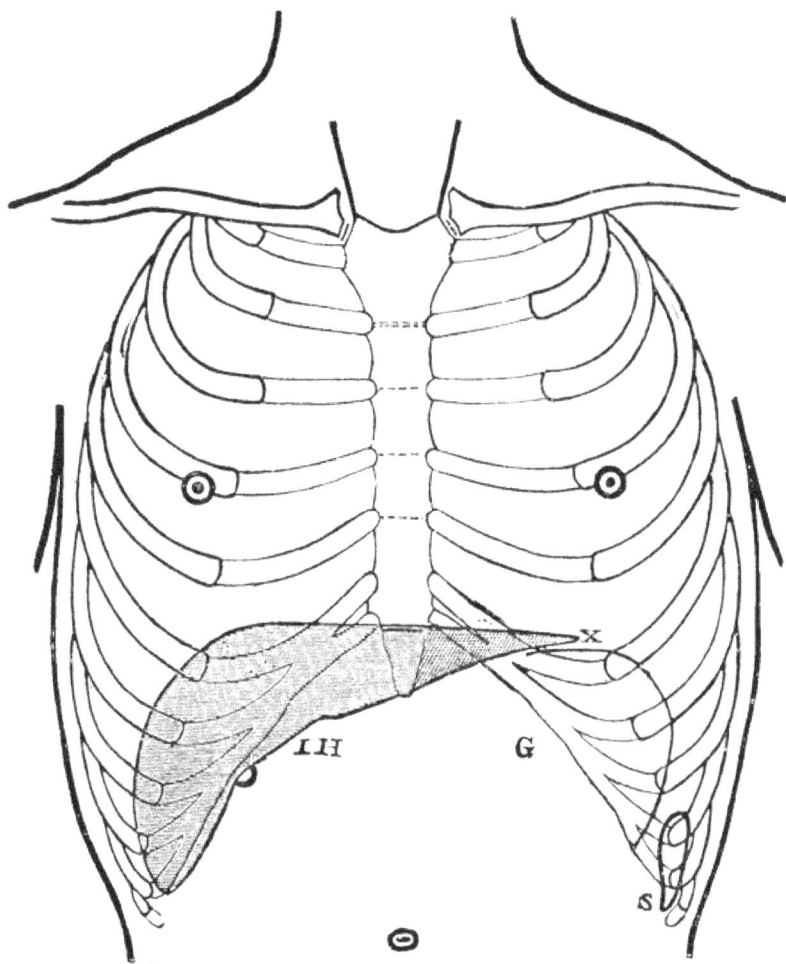

This Outline should be compared with the following Outlines belonging to the same series. IIX—The upper limit of absolute dulness of the liver. III—The inferior hepatic border.

SECOND STEP IN CARDIAC PERCUSSION:

HOW TO DETERMINE THE ABSOLUTE CARDIAC DULNESS.

(1) The lower boundary we have already secured, since it forms part of the hepatic line HX.

(2) In order to find the right boundary, place the finger to be percussed vertically in the 5th interspace, one and a half inch to the left of the sternum : the note struck here will be absolutely dull ;

(3) Now percuss over the middle of the sternum. This will be much less dull, perhaps even resonant ; and a boundary will be readily traced between the dull and the resonant parts, coinciding with the left sternal border.

(4) This line of dulness will be found to extend along the left edge of the sternum, from the lower border of the 4th cartilage to the hepatic line.

(5) In seeking to determine the left boundary, we should compare the percussion note of the dull area, as far as made out hitherto, with the full resonance of the left 3d and 4th interspaces, outside the nipple line,—neglecting for the present the partial dulness which may be detected in the left outer and upper part of the præcordium. The observer will easily work out the oblique boundary line depicted in the diagram by alternately percussing the dull and the resonant surfaces.

The present Outline, shewing the combined areas of absolute dulness, cardiac and hepatic, illustrates the results arrived at so far. The dull area of the heart and that of the liver have a boundary in common, viz.: the hepatic line. We have assumed that this line followed a straight course, but of this we have not supplied evidence. In other words, that portion of the hepatic line which extends between the infrasternal notch and the cardiac apex has not been determined by percussion, but in an indirect way. This is a weak point which will presently claim further attention.

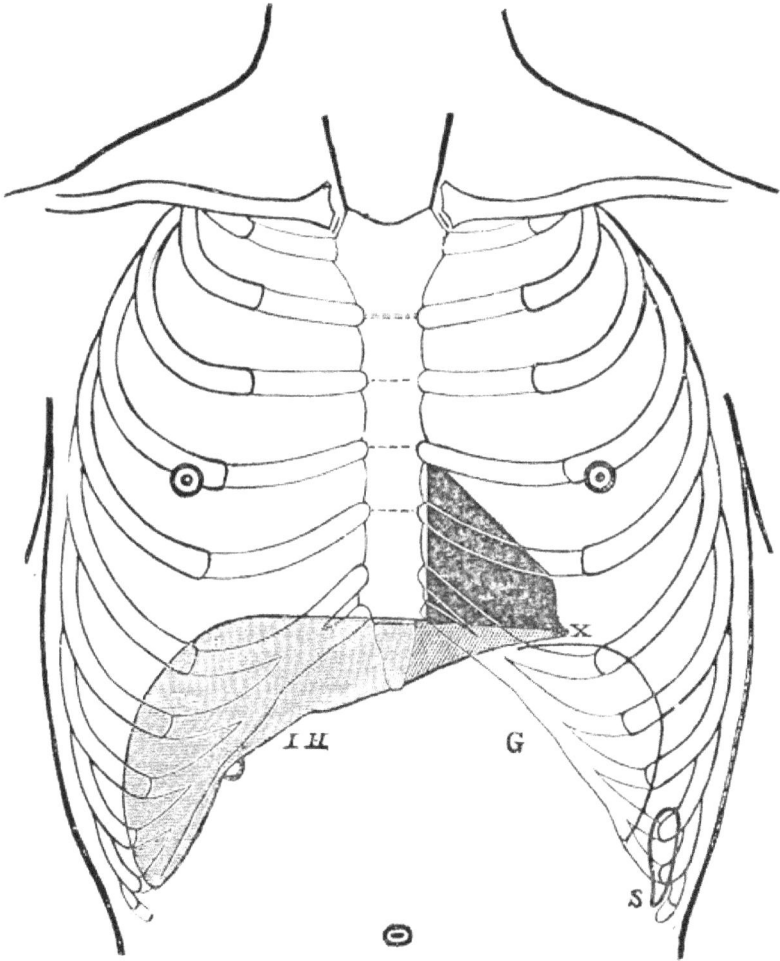

Compare Outlines XXV. and XXVI. The dulness of the two areas is
continuous in the left chest, *although not of identical quality.* The left
portion of the hepatic line is almost occluded.

THE RESPIRATORY VARIATIONS OF THE LEFT OR UPPER BOUNDARY OF ABSOLUTE CARDIAC DULNESS.

We have hitherto spoken of the boundaries of the absolutely dull area as though they were constant, at least in the same individual. This is not the case. They all vary with respiration, but the variation is noticeable only in the left upper boundary, and in the hepatic line.

I. So long as no pleural or pericardial adhesions exist, the effect of a deep inspiration is to cause the left upper lobe of the lung to expand down to a lower limit than previously. Its inferior border will cover from above more of the cardiac surface ; and the absolute dulness will be reduced by so much. Conversely, the next expiration will correct this encroachment. Therefore, held, forcible expirations may lead to a more than average size of the dull area. These changes will not result if either the upper lobe of the left lung, or the anterior surface of the heart be closely united by adhesions to the anterior chest-walls.

II. Meanwhile, unless the heart be closely adherent to the pericardium and to the chest-wall, a forcible inspiration will depress the central portion of the diaphragm by an appreciable amount. The dull area, encroached upon from above, will thus tend to be increased downwards. During forcible expiration the converse change will take place. But the range of movement of the central tendon of the diaphragm is much more restricted than that of the pulmonary fringe ; and the variations in the size of the dull area will be governed by the movements of the latter.

From these facts the following conclusions may be drawn :

(1) The size of the area of absolute dulness is in health not fixed but variable.

(2) It increases with each expiration, and lessens with each inspiration, proportionately to the depth of the breathing.

(3) The mobility of the oblique line (or left boundary) of dulness is, in itself, evidence that close adhesions do not exist between the left upper lobe and the chest wall, nor between the latter and the front of the heart.

(4) A determination, in individual cases, of the presence or absence of the respiratory variations in question, is, therefore, practically useful for purposes of diagnosis.

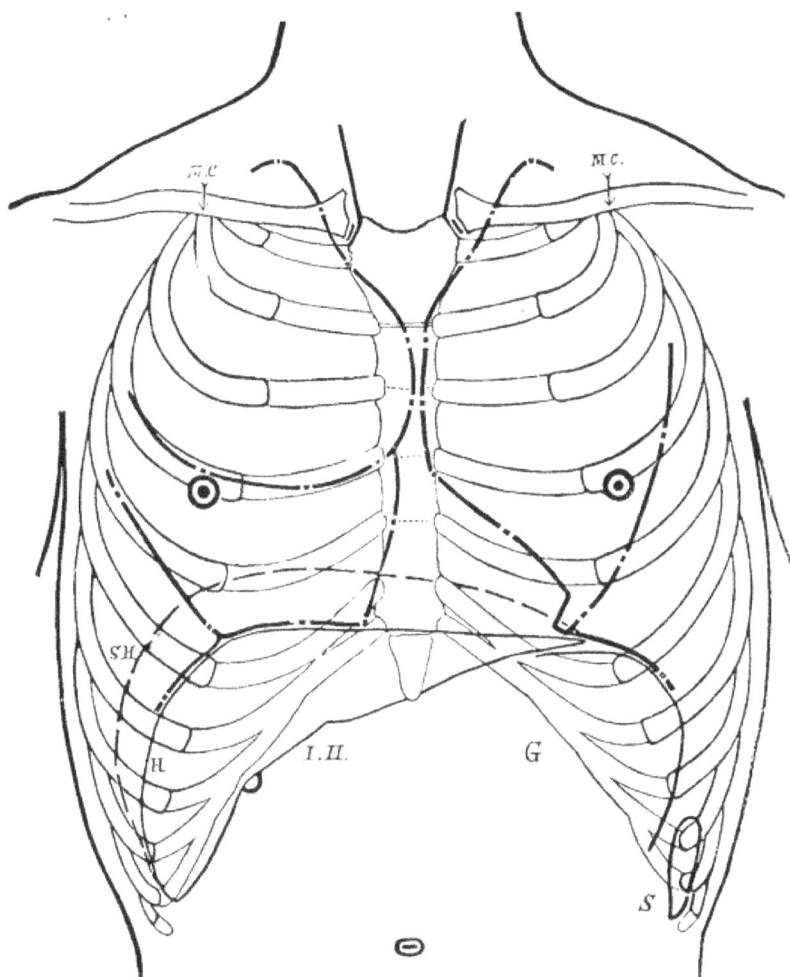

The anterior edges of the lungs are shown in interrupted black line. The lungs are supposed to be rather fully inflated. During forced expiration the oblique lower border of the left upper lobe would approach the level of the fourth cartilage.

THIRD STEP IN CARDIAC PERCUSSION:

HOW TO DETERMINE THE PARTIAL DULNESS OF THE LIVER.

The hepatic line HX, shewn in the Outlines, does not represent the highest level of the liver within the thorax, but only that level below which the liver is, and above which it is not, in contact with the anterior thoracic wall. Our present business is to determine the upper hepatic boundary. The interval between the chest wall and the receding hepatic surface being normally occupied by lung, we shall expect any dulness due to the liver to be toned down by so much pulmonary resonance. This modification is best described by the term "partial dulness" (in preference to "deep dulness," which is an ambiguous expression).

Percussors usually determine the upper hepatic boundary in the inner inframammary region; and this is, in every way, the most favourable situation to select for this purpose.

The boundary sought will be found without difficulty by alternately percussing the right 4th and 5th interspaces, between the nipple and the parasternal lines. In the Outline it occupies the level of the 5th cartilage; but the student will bear in mind that, as stated above, its position is a varying one. Each deep inspiration will cause a perceptible descent of the liver, which will rise again with expiration.

For ordinary purposes this simple determination of level will suffice. The student, however, should complete the examination by percussing out the boundary of partial dulness towards the right extremity of the liver, as shewn in the present Outline, and towards the left, as will be described further on. The boundary in question is seen to have a nearly horizontal portion, that nearer the middle line; and a curved descending portion, that extending towards the axillary line.

The horizontal portion may conveniently be termed "the suprahepatic line." This name, which the author has ventured to introduce, conveys its own explanation. Taken in conjunction with the equally clear expressions, "infrahepatic line," or "lower hepatic border," it illustrates by contrast the meaning of the term, "hepatic line," previously described.

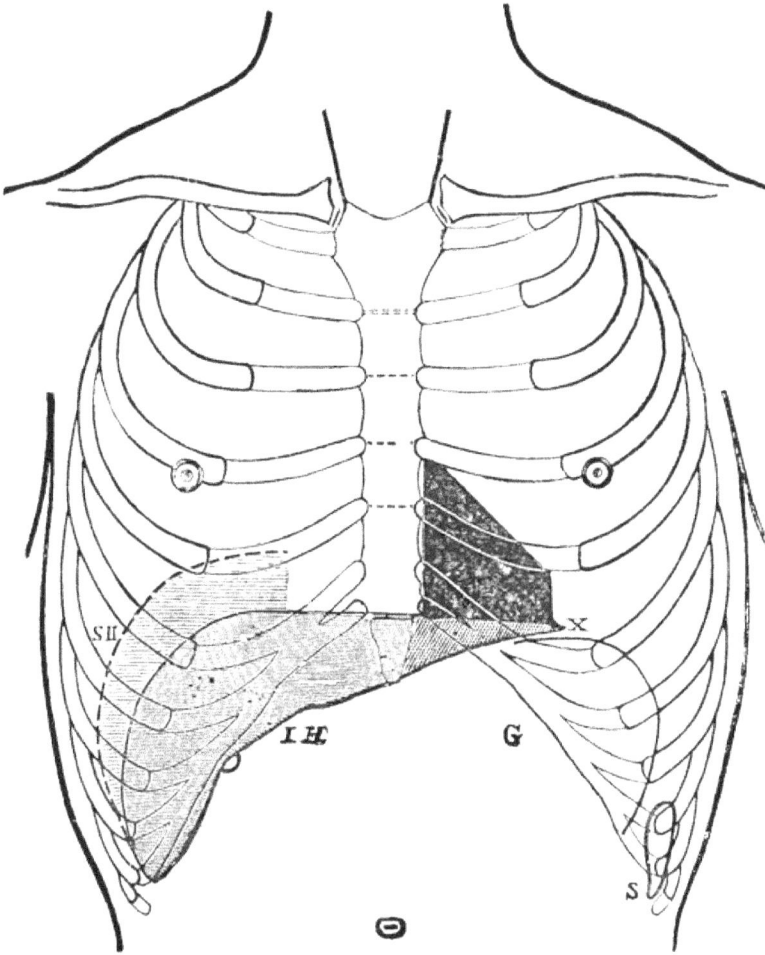

HX—The hepatic line, or upper boundary of absolute hepatic dulness.
SH—The suprahepatic line, or upper boundary of the liver, and of the partial
hepatic dulness.   IH—The infrahepatic line, or lower border of the liver.

FOURTH STEP IN CARDIAC PERCUSSION.

HOW TO DETERMINE THE PARTIAL CARDIAC DULNESS IN THE RIGHT HALF OF THE CHEST.

In this somewhat more difficult portion of the examination we must chiefly rely upon the fairly sharp contrast given by the fully resonant lung in the mammary region. We have not in this case any equivalent degree of cardiac dulness to compare with the full pulmonary resonance. The sternal portion of the præcordium is resonant, and the right parasternal portion is hardly ever quite dull. Nevertheless, when percussing from left to right, the trained percussor (especially if using the pleximeter) will not miss the line at which the partially resonant note suddenly acquires unmixed clearness as he passes beyond the extreme right cardiac boundary.

In order successfully to conduct this examination the finger to be percussed must be moved from the horizontal into the vertical position ; this will facilitate a determination of the vertical line shewn in the diagram.

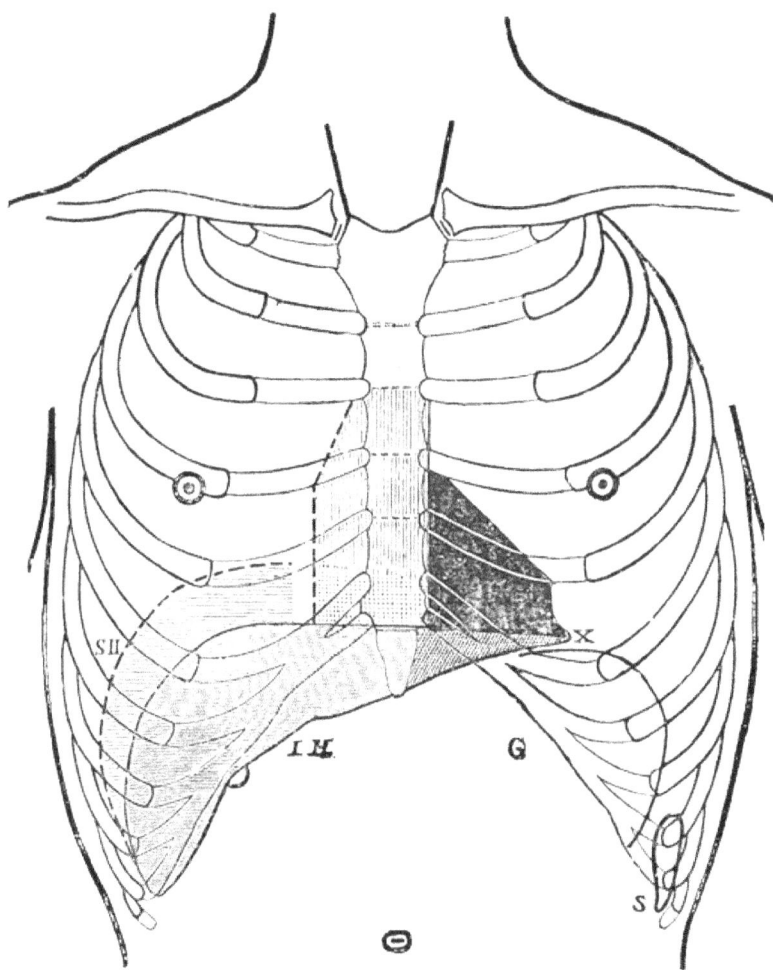

HX—The hepatic line. SH—The suprahepatic line. The shaded portion of the sternal and right parasternal regions represents partial dulness. The space left blank has not yet been carefully percussed.

FIFTH STEP IN CARDIAC PERCUSSION.

HOW TO DETERMINE THE PARTIAL CARDIAC DULNESS IN THE LEFT HALF OF THE CHEST.

The left boundary of partial dulness is rather more difficult to find than the right. In attempting this percussion it is therefore advisable to consult at first the normal cardiac Outline. Here again as in the preceding stage, the fulness of the pulmonary resonance with which the partial dulness has to be compared renders the task somewhat easier. In the Outline the oblique boundary is seen to follow the line joining the 3d left chondrosternal junction to the 4th left chondrocostal junction.

The shorter vertical boundary, which extends from the left nipple level to the hepatic line, is often somewhat obscured by conducted resonance due to the stomach. This occurs almost invariably at the lower extremity of the line : and the surface corresponding to the apex-beat is found to be relatively resonant, instead of belonging, as it otherwise would, to the area of absolute dulness, or at least to the partly dull area.

In the female a determination of the left area of partial dulness is difficult and apt to be painful, and should not be insisted on as a matter of routine.

N. B. Within the region of partial dulness the student may often observe more or less definite differences in the note of percussion. He should not be led away by these less important distinctions from the broad lines which have been described. Mention may, however, be made here of the two accessory lines of incipient dulness seen in Outline XXXII., one at the right anterior thoracic base, the other at the left upper thoracic region. They are not practically of importance to the beginner.

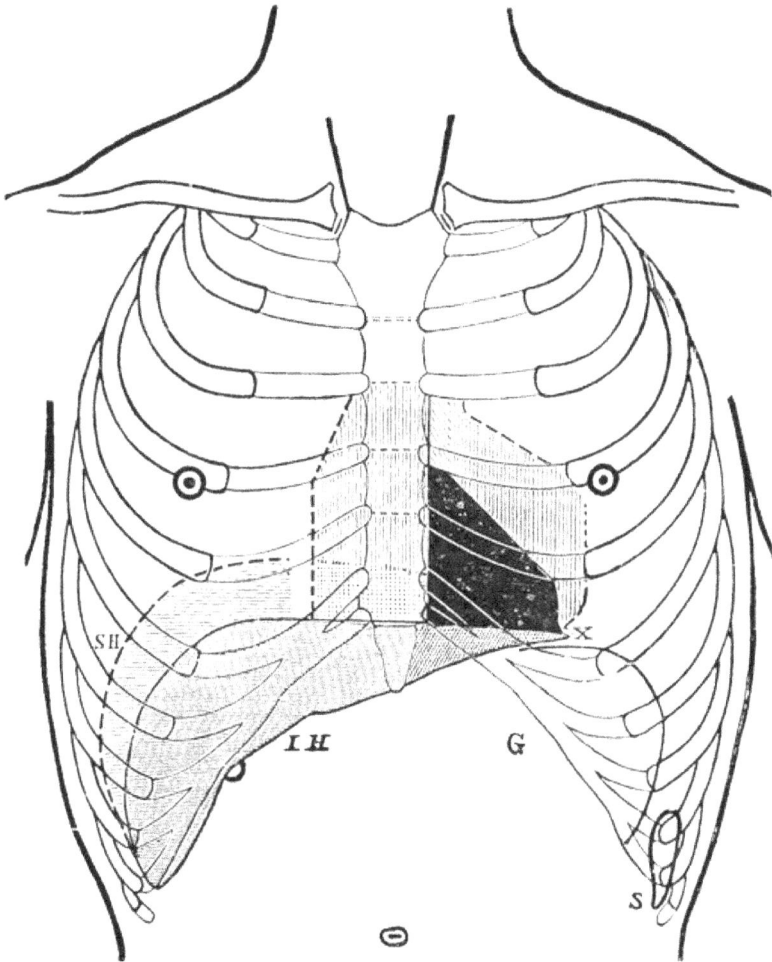

HX—The hepatic line. SH—The suprahepatic line. The shaded surface of partial dulness in the left chest is bounded above by an oblique line and to the left by a vertical line, which are described on the opposite page.

SIXTH STEP IN CARDIAC PERCUSSION :

HOW TO DETERMINE THE PRÆVASCULAR DULNESS.

Careful percussion conducted across from the right to the left parasternal
line will not fail to detect variations in the degree of resonance. No part
of this surface is absolutely dull, although, as a whole, it is noticeably less
resonant than the neighbouring regions. This change is due

    (1) Partly to the increasing thinness of the pulmonary fringe,

    (2) Partly to the density of the mediastinal contents (mainly vessels,
        hence the expression " prævascular dulness "), and

    (3) Partly also to the Sternum.

THE STERNAL NOTE.

The Sternum has already been described as a conductor of the sonorous
pulmonary vibrations. Equally well must it conduct the less ample vibrations described as " high-pitched " or " dull," which arise from denser
structures in contact with it. From this conflict or combination of vibrations results an " average sternal note," which in the upper chest is less
resonant, in the lower chest more resonant, than the note elicited on either
side of the bone, and which enables us to define by percussion the shape of
the sternum. Indeed we cannot avoid this determination ; it forms part
of every percussion conducted over the front of the chest. The absolute
cardiac dulness, below, owes its straight inner outline to the sternal note.
In the same manner the normal prævascular dulness, above, assumes the
shape depicted in this Outline because this is the shape of the underlying
sternum and manubrium. It is relatively narrow below, and broadens
upwards.

The prævascular dulness can sometimes be traced on either side beyond
the sternum, even in health. In disease this extension may become very
marked. Usually, whether in health or in disease, it is not as great on the
right as it is on the left side. A narrow vertical strip of dulness ($\frac{1}{4}$ to
$\frac{1}{2}$ inch wide) extending along the left sternal border, can be recognised by
percussion in almost every instance. It is shewn in this Outline. The

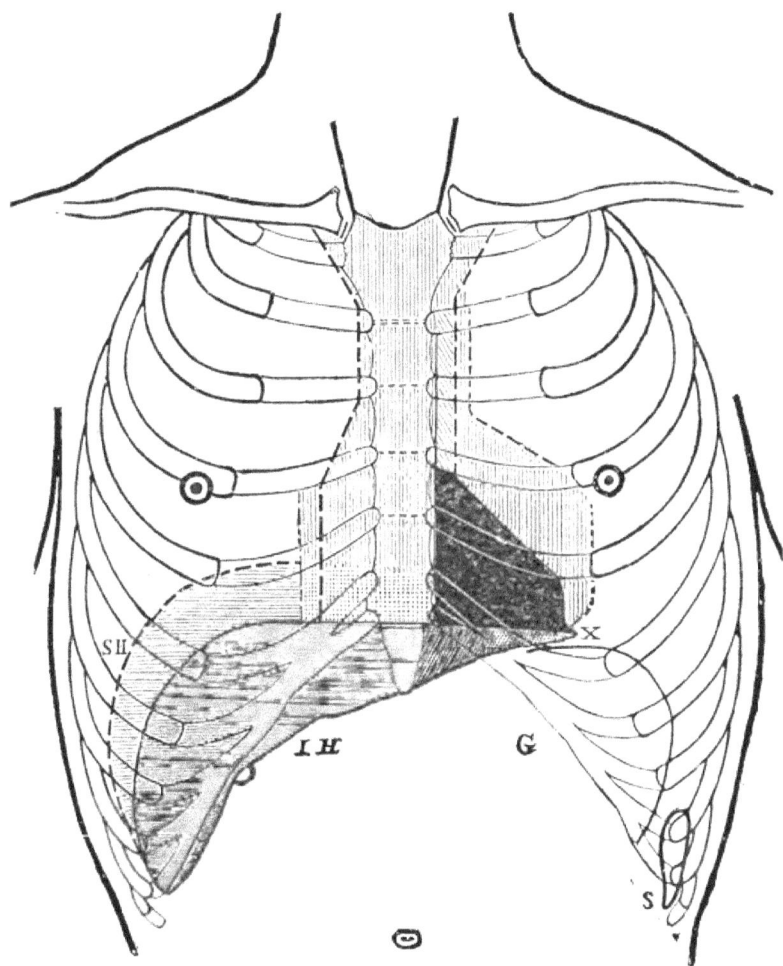

A narrow strip is seen to the left side of the sternum, continuing upwards the outline of the absolutely dull area. This strip is duller than the surfaces between which it is placed, but not absolutely dull. The sternal surface yields partial dulness only, throughout.

strip in question is normally duller than the sternum ; and it is found to be continuous with the area of absolute cardiac dulness. For this concordance of lines (which is invariably found by careful percussion) good anatomical reasons exist.*

The additional line on the left of the prævascular dulness, as well as the additional line seen in this diagram to the right of the cardiac dulness are both easily discovered. The surfaces which they define are areas of subresonance, rather than of partial dulness. Sansom's pleximeter will be found very useful in tracing these boundaries, as well as in connection with the final stage of cardiac percussion.

---

* The same reasons explain the striking concordance which a good percussor invariably finds between the lines which he has traced from above and those traced from below.

SEVENTH AND FINAL STEP IN CARDIAC PER-
CUSSION :

THE HEPATIC AND THE SUPRAHEPATIC LINES AS OBTAINED BY PERCUSSION.

We are competent to find by percussion both the hepatic and the supra-hepatic line, not only in the right chest, but also in the left. In other words, we can percuss out with accuracy, on the anterior surface of the body, the whole outline of the liver on the one hand, and the whole outline of the heart on the other.*

Moreover, we can prove by percussion that these two organs overlap from front to back ; and we are able to trace the extent of their overlapping. This is easily accomplished, as soon as their anatomical relations, which have not always been sufficiently regarded, are thoroughly understood. In the ordinary routine of physical examinations of the heart the lines in ques-tion are best traced at an earlier stage ; a different course has been adopted here to facilitate their demonstration.

I. THE HEPATIC LINE.

Let it be stated once more that the *Suprahepatic line* marks the highest level reached by the liver in the depth of the chest ; whilst the *Hepatic line*

---

* There is no reason why an experienced percussor should not succeed in this with un-aided finger percussion. Nevertheless Sansom's pleximeter is a great help and it was with its assistance that the author first clearly defined the two lines in question (see *Lancet*, Aug. 29, 1891).

indicates the highest level of actual contact between the liver and the anterior chest-wall ; this being, at the same time, the lowest level reached by the heart. A drawing representing merely the parts situated immediately behind the anterior chest-wall would shew the hepatic line, but it would not shew the suprahepatic line ; it might shew the lowest level of the heart and of the peri-cardial floor ; but it could not shew the highest level of the latter.

The annexed illustration is of this kind. In it only the foremost intrathoracic plane is shewn. The suprahepatic line does not therefore come into view.

The liver and the heart being not only in contact, but partly wedged one in front of the other, and both being dense organs, it might have been very difficult (and is still most often alleged to be impossible) to trace their mutual boundary by percussion. But the effect of the gastric resonance is to transform, in the left half of the chest, the hepatic dulness into a modified hepatic reso-nance. Thanks to this circum-stance all difficulty is removed.

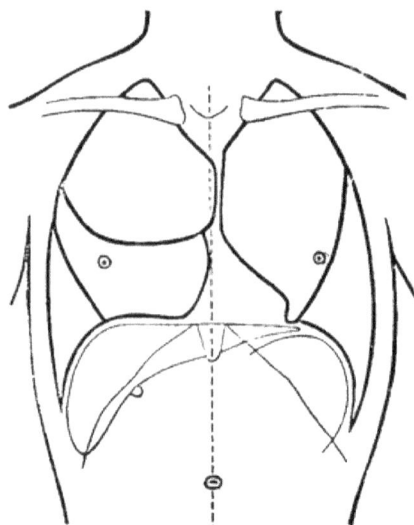

FIG. XXXIII.

THE HEPATIC LINE IN ITS RELATIONS TO THE LUNGS, INFRASTERNAL ANGLE, AND STOMACH.

A simple diagram will shew more clearly than words can explain how the hepatic line, or anterior cardio-hepatic boundary, can be determined by the ordinary method of contrast in percussion.

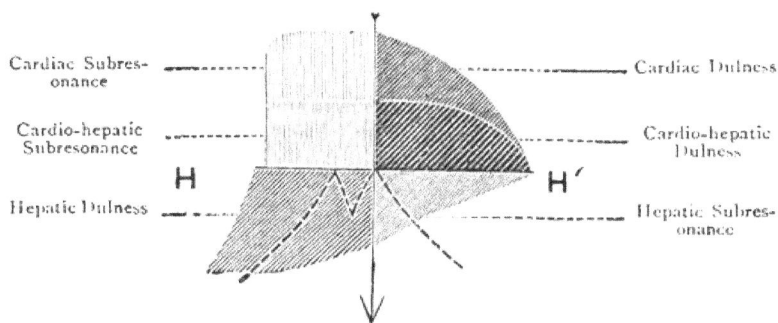

Cardiac Subres-
onance

Cardio-hepatic
Subresonance

Hepatic Dulness

H   H′

Cardiac Dulness

Cardio-hepatic
Dulness

Hepatic Subres-
onance

FIG. XXXIV.

Diagram representing the hepatic line HH′ and the relative qualities of
the sounds obtained by percussing above and below the line. Part of the
costal arch, the infrasternal angle, and the xiphoid cartilage are represented
in interrupted line. The arrow passing through the left chondroxiphoid
angle represents the left edge of the sternum and prolongation of the
same line downwards.

The diagram shews that, whereas in the right chest the hepatic surface
below the line HH′ is dull, and the cardiac surface immediately above it,
comparatively resonant, in the left chest the hepatic surface is comparatively resonant and the cardiac surface above it is absolutely dull. We may
therefore feel quite certain that percussion will avail to mark off the lower
boundary of the cardiac dulness from the dulness of the liver ; and in view
of the clinical advantage which we may secure, we should not neglect to put
to the test our own ability to carry out this method.

II. THE SUPRAHEPATIC LINE.

(The following remarks will be more readily understood in conjunction
with Outline VIII. which shews diagrammatically the upper surface of
the liver and the pericardial floor, the heart having been removed. It

will be seen that the deep-seated convexity of the liver extends across the chest as far as the left extremity of the organ, and is parallel with the hepatic line, except at this left extremity, where the liver becomes reduced in thickness.)

In the right inframammary region the pulmonary resonance was found to be impaired at the level of the 5th cartilage, and this was ascribed to the presence, behind a layer of lung, of the deep convexity of the liver.

Precisely the same influence might be expected to obtain farther toward the left, and it does make itself felt there. Let us consider first the partial cardiac dulness over the lower third of the sternum.

In this situation, although the heart is superficial, yet, owing to resonance conducted by bone from a distance, the dulness is not absolute. But where the deep convexity of the liver passes behind the lower part of the præcordium, the presence of this organ should be indicated by an increase of the dulness ; and if percussion of the sternum be practised from above downwards, a transverse boundary should be found in the cardiac area between the more resonant upper part or cardiac subresonance, and the less resonant lower part or cardio-hepatic subresonance. This is precisely the result obtained by careful percussion ; and the line will prove to be an exact continuation of the line SII (see Outline XXXV).

Passing now to the area of absolute cardiac dulness, a difference in the degree of dulness should likewise be recognisable between its upper and its lower district, the lower or cardio-hepatic area being more deeply dull by reason of the combined cardiac and hepatic dulnesses. Outline XXXV. shews this distinction in a very accentuated manner. With the help of a Sansom's pleximeter convincing proof of its reality can be obtained by any observer.

Having once been recognised, the line of demarcation is easily found again, either in the same subject or in others. After having sought it according to the easier method, the student will probably succeed in finding it at once in the middle of the area of absolute cardiac dulness, and in tracing it back from left to right, until it merges into the upper line of hepatic dulness, originally defined in the right inframammary region.

This concludes the description of the method and of its normal results. The ensuing pages will give some insight into the application of the method to pathology.

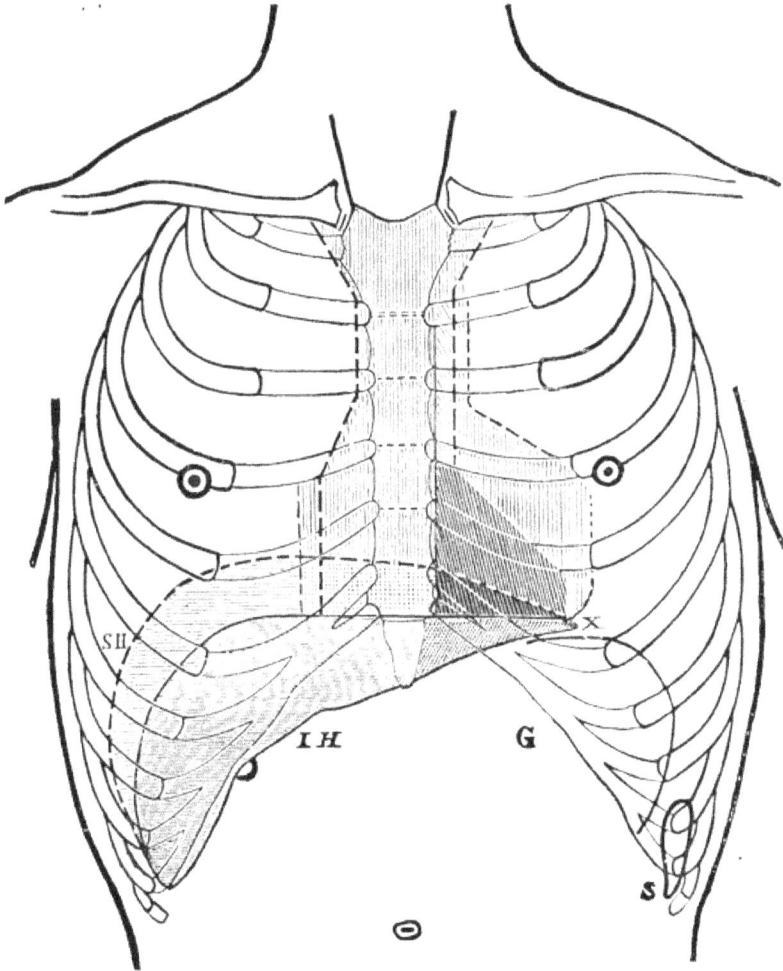

FIG. XXXV.

SH—The suprahepatic line, separating, on the right side, subresonance above from partial dulness below, and, on the left side, absolute dulness of less degree above from that of greater degree below. HX—The hepatic line, below which—absolute hepatic dulness on the right, and modified hepatic resonance on the left.

CARDIAC PERCUSSION IN ABNORMAL CONDITIONS.

The value of an accurate knowledge of the position of the normal lines of percussion will become apparent to the clinical student in connection with the displacements to which they are liable in disease.

I.

THE PATHOLOGICAL VARIATIONS IN THE POSITION OF THE HEPATIC AND OF THE SUPRAHEPATIC LINES.

(1) Any abdominal influence occasioning a uniform rise of the diaphragm, will raise both the hepatic and the suprahepatic line above their average level.

(2) Any thoracic influence occasioning a uniform depression of the diaphragm will lower both these lines.

(3) Any influence either depressing or raising the diaphragm on one side only will tilt one end of the liver or the other, and occasion an obliquity of the two lines.

(4) Lastly, any alteration in the vertical thickness of the liver will cause the two lines to be either nearer to each other, or farther apart.

II.

THE PATHOLOGICAL VARIATIONS IN THE LEVEL OF THE HEART, AND THEIR CAUSES.

The practical meaning of the preceding statements is that any variation in the level of the liver as a whole will likewise affect the level of the heart; and that, conversely, any rise or fall in the lower cardiac level will be bound up with a similar rise or fall of the subjacent hepatic surface; the usual, although not invariable, result being a change in the direction of the longitudinal axis of the liver.

An unusually high position of the heart is almost always of pulmonary rather than of cardiac origin; cardiac atrophy may, however, be an occasional cause. On the other hand, the causes of a displacement of the heart downwards usually reside in the pericardium or in the heart itself, and depend upon variations in their volume.

Roughly speaking, *an increase in the size of the heart* may take place in one of two directions :

> The heart may enlarge in its transverse axis ;
> It may enlarge vertically.

The first of these changes will be recognised by a shifting of the lateral boundaries. The second will lead to a depression not only of the lower cardiac boundary but also of the left extremity of the hepatic line. The natural declivity of the liver downwards and to the left will then be increased. This is one of the most striking results of left cardiac hypertrophy, whenever it attains considerable proportions.

Enlargement of the pericardium by fluid effusions invariably depresses the liver. *Cæteris paribus* the depression is greater on the left side than on the right, and the hepatic line assumes an obliquity analogous to that which results from left cardiac hypertrophy.

CARDIAC HYPERTROPHY :

ITS VARIETIES AND ITS INFLUENCE ON CARDIAC DULNESS.

Cardiac Hypertrophy may be general or partial.

If partial, cardiac hypertrophy may affect the Right or the Left side of the heart.

I.

GENERAL CARDIAC HYPERTROPHY.

(1) If due to healthy increase in the muscular and respiratory functions, the hypertrophy will seldom be great. An increase in total cardiac dulness may be perceived ; but this will not be considerable, and the liver will not be greatly depressed nor the cardiac apex moved far outwards.

(2) If, on the contrary, it be due to adherent pericardium, hypertrophy may attain a great size, and will make itself known

> (1) by much increased absolute dulness,
> (2) by outward and downward displacement of the apex beat, and
> (3) by downward displacement of the liver.

II.

PARTIAL HYPERTROPHY LIMITED TO THE RIGHT VENTRICLE.

When not complicated with dilatation, this does not appreciably modify the anatomical lines of percussion. Hypertrophy complicated with dilatation will be mentioned farther on.

III.

PARTIAL HYPERTROPHY LIMITED TO THE LEFT VENTRICLE.

This gives rise to very definite results easily recognised. The apex beat is

(1) depressed,

(2) displaced towards the left, and

(3) at the same time very forcible.

The left extremity of the hepatic line is apt to be depressed (this change should be looked for only in cases of extreme hypertrophy) and may thus become oblique downwards and towards the left in a very marked degree.

CARDIAC DILATATION :

ITS VARIETIES AND ITS INFLUENCE ON CARDIAC DULNESS.

The subject of cardiac dilatation might be subdivided into the following groups :

I. General dilatation of the heart,
- without hypertrophy.
- with hypertrophy
  - general
  - or
  - partial.

II. Partial dilatation of the heart,
- without hypertrophy.
- with hypertrophy
  - of the whole heart.
  - of the dilated side only.
  - of the other side only.

Since, however, the extent and the direction of any increase in dulness due to this cause vary according to the portion of the heart which is affected, we shall adopt the following subdivision ;

I. General dilatation,

II. Partial dilatation
$\begin{cases} \text{(A) limited to the right auricle.} \\ \text{(B) limited to the right ventricle.} \\ \text{(C) limited to the left auricle.} \\ \text{(D) limited to the left ventricle.} \end{cases}$

I.
GENERAL DILATATION.

Seldom primary, except as a result of anæmia, general dilatation often comes under observation at a late stage of cardiac disease, when this has ceased to be complicated with much hypertrophy. Hearts in this condition are usually very bulky, either from the dilatation being general, or because one side still retains some compensatory hypertrophy.

II.
PARTIAL DILATATION.

A.—DILATATION LIMITED TO THE RIGHT AURICLE.

This dilatation invariably throws the extreme right cardiac boundary farther outwards—not always, however, in proportion to the increased bulk, since some displacement towards the left may occur as a result of other circumstances.

B.—DILATATION LIMITED TO THE RIGHT VENTRICLE.

Dilatation of the right ventricle invariably throws the left heart, unless bound down, farther outwards, and the liver slightly downwards. The cardiac displacement predominates over the hepatic so long as the heart remains, as in health, freely movable. Fig. XXXVI. shews that the right ventricle can expand with freedom only in these two directions, its antero-posterior diameter being confined by unyielding structures.

C.—DILATATION LIMITED TO THE LEFT AURICLE.

Dilatation of the left auricle, if considerable, gives rise to pulmonary pressure symptoms, but not to any physical signs, except downward displacement of the apex beat.

D.—DILATATION LIMITED TO THE LEFT VENTRICLE.

This dilatation markedly depresses the apex beat, with displacement outwards. Here, as in c, the left extremity of the hepatic line may be alone or chiefly displaced,

THE ENCROACHMENT ON THORACIC SPACE DUE TO COMBINED GENERAL HYPERTROPHY AND DILATATION OF THE HEART.

Sometimes the heart is much hypertrophied as well as much dilated. When the change is general (as in some instances of adherent pericardium at a late stage), the heart may be very large. The absolute dulness is then greatly increased above and at the sides, and the diaphragm and the liver may be markedly depressed. The resulting encroachment on thoracic space is specially noticeable in the relatively less capacious chest of children. This feature is well shewn in the present Outline.

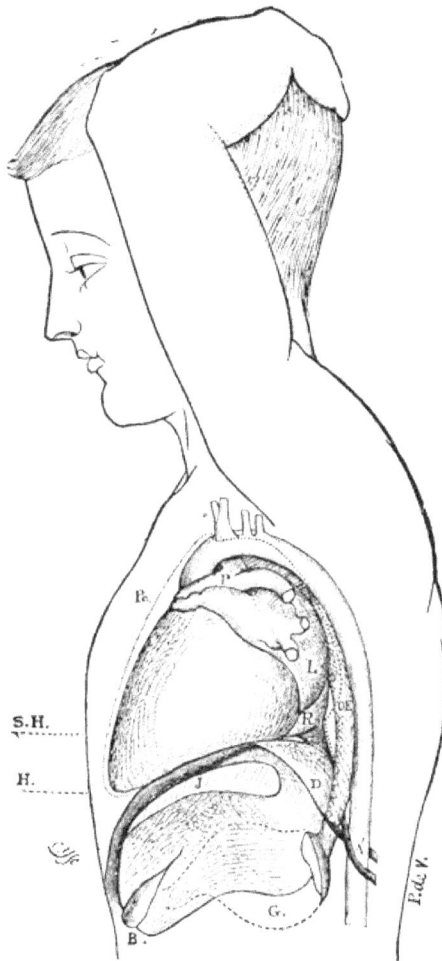

FIG. XXXVI.

The heart and liver are seen from the left side ; a vertical antero-posterior section is supposed to have been made immediately to the left of the heart. A—Aorta. B—Gall bladder. D—Diaphragm. G—Outline of stomach. J—Vertical section through the left extremity of the liver. H—Level of the hepatic line. SH—Level of the suprahepatic line. P—Pulmonary artery. Pc—Outline of the pericardium, shewing the line of reflection from the great vessels. L—Left auricle. Œ—Œsophagus. R—Right auricle and Vena Cava.

THE CHANGES IN THE RELATIONS OF PARTS IN PERICARDIAL EFFUSION ; AND THE DIAGNOSIS OF EFFUSION BY PERCUSSION.

Percussion furnishes the positive element in the diagnosis, viz., increase in the area of præcordial dulness ; palpation and auscultation furnish the negative elements, which are diminution or loss of heart-beat and of heart-sounds. The methods of palpation and of auscultation are described under the corresponding headings.

The changes in the præcordial dulness discovered by percussion are :

Alterations in shape.
Alterations in size.
Alterations in quality.

Their extent will vary with the amount of fluid effused, the recognition of a moderately large effusion presenting, as a rule, no difficulty.

(1) When percussed, the pericardial enlargement is found to be of a typical shape. The dull area is broad at the base and tapers more or less evenly towards the manubrium sterni.

In considerable effusion the peaked summit will be exchanged for a more or less spherical outline of the upper border of dulness ; the lower border remaining, as before, rectilinear. Outline XXXVII. represents a stage between the peaked and the globular form.

(2) The alteration in the extent of absolute dulness is apt to be undervalued owing to the conducted resonance of the sternum and to the overlapping of the fringe of the lung in front of the effusion ; but some dulness always extends at least as far as the right sternal border.

(3) The quality of the dull note is characteristic. There is " deadness " or complete absence of elastic as well as of sonorous vibration over the dull area. This sign is an important help in all doubtful cases.

The Outline, which has been constructed on the basis of clinical cases observed, displays :

The shape of the distended pericardium :

(a) This is always broader at the base,
(b) In small effusions, pyriform,—the normal shape of the præcordium exaggerated ;
(c) In large effusions almost hemispherical.

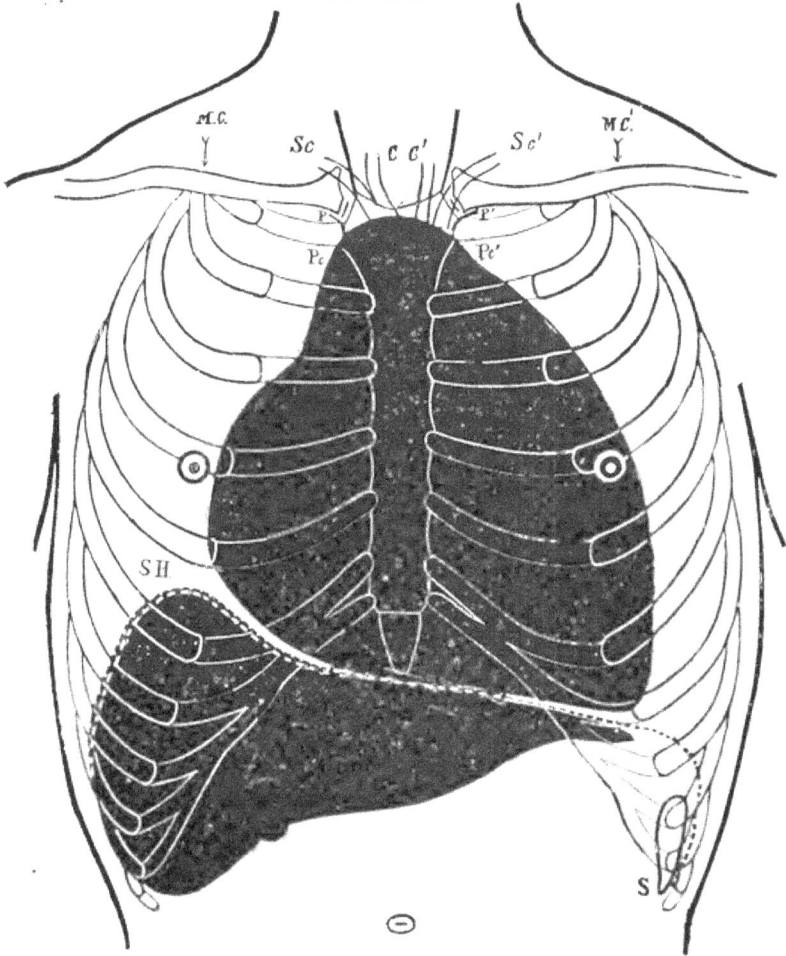

FIG XXXVII

Pe and Pe'—Right and left pleuro-pericardial layer, distended by fluid. P and P'—Pleural layer continued upwards from the pleuro-pericardium. Sc and Sc'—Right and left subclavian artery. C and C'—Right and left carotid. SH—Upper surface of the liver, depressed and moulded by the weight of the distended sac.

The predominance of pericardial dilatation towards the left. This
usually causes a bulging of the left side of the chest, which is
not shewn here.

The much increased vertical diameter.

The downward bulge of the floor of the pericardium.

The depression of the liver.

The symmetrical depression of the diaphragm,—causing depression of the stomach, and, to a less extent, of the spleen.

The great increase in the absolute præcordial dulness.

The slighter increase in the absolute hepatic dulness (which is displaced).

The much extended line of contact between præcordial and hepatic
dulnesses, which are nevertheless perfectly distinguishable from
one another by percussion.

THE HEART IN PERICARDIAL DROPSY.

This Outline is based upon measurements taken after death in a case of
Pericarditis.[1] The heart is displayed as it lies in the middle of the effusion.
The points specially to be noted are :

(1) The distension of the pericardial sac ;

(2) The central and isolated position of the heart surrounded by fluid ;

(3) The interval between the lower anterior border of the heart and the
     floor of the pericardium, Pc ;

(4) The stretching of the heart and of the large vessels (especially in the
     orthopnœic posture, which is the rule in large effusions) ;

(5) The depression of the floor of the pericardium (in this case, as far as
     the level of the tip of the xiphoid cartilage) ;

(6) The depression of the liver to the same extent ;

(7) Lastly, not so well shewn as in the Outline XXXVII., the lateral
     bulging of the lower part of the sac on either side of its attach-
     ment to the midriff.

It will be noticed that a cardiac impulse perceived, say in the 5th interspace, would not coincide, as in the normal subject, with the lower limit
of præcordial dulness, but would be decidedly higher than that line.

[1] A sketch of the pericardium *in situ* was taken by Dr. H. B. Grimsdale, whose
valuable assistance the author is happy to acknowledge.

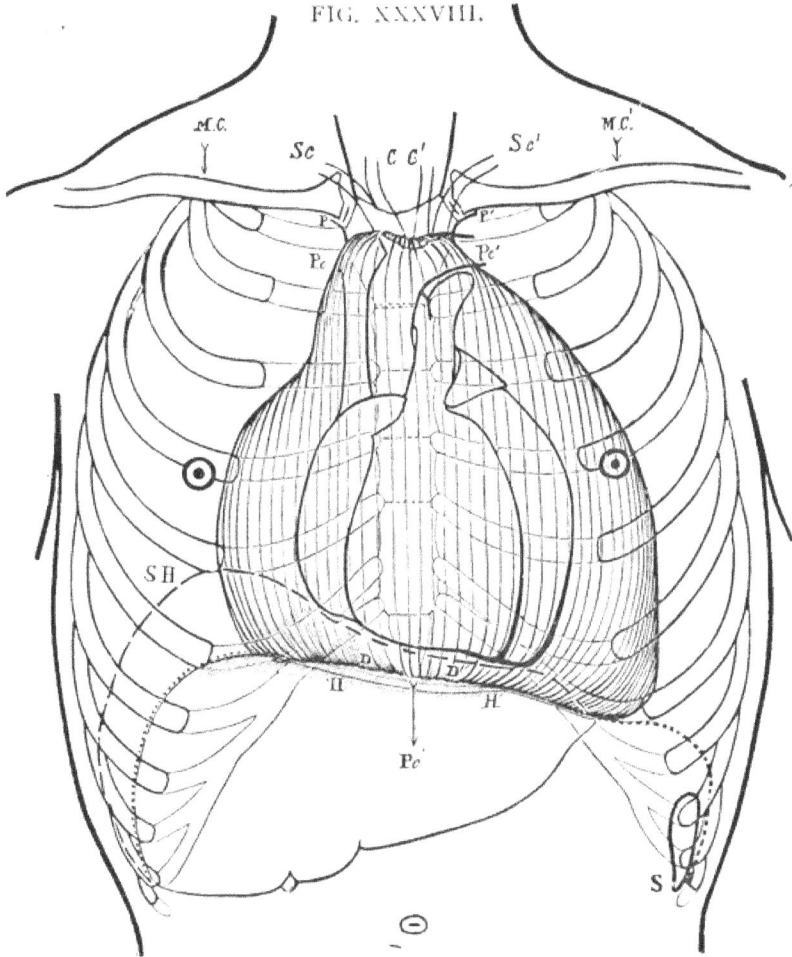

FIG. XXXVIII.

The shaded surface represents the distended pericardium, imagined to be transparent and to allow the outline of the heart to be seen. Most of the letters have the same meaning as in Outline XXXVII. SH—The supra-hepatic line. HH'—The hepatic line, and DD' the floor of the pericardium, normally contiguous with, but here separated by fluid from, the anterior lower border of the heart. At Pc' the floor of the pericardium is seen to be depressed, at its point of greatest convexity downwards, as far as the level of the extremity of the xiphoid cartilage.

PART V.

CARDIAC AUSCULTATION IN THE NORMAL SUBJECT AND IN PATHOLOGICAL CONDITIONS.

It is essential that the beginner in cardiac auscultation should refresh his memory as to the anatomy of the heart, and particularly of the heart valves. The next few illustrations are devoted to this object. The two first shew the relation of the valves to each other; those that follow deal with their relation to the anterior chest wall, and with their topography.

The Clinical Outlines belonging to this series deal, in the first place, with the normal heart-sounds and with the situations in which they are severally audible; and inasmuch as a graphic representation of the heart-sounds is sometimes desirable, a code of suitable symbols has been devised.

The abnormal sounds or " murmurs " may also with advantage be expressed by symbols, and a set of these is suggested for use at the bedside.

Separate Outlines are devoted to the description of the several murmurs with reference to their site of intensity and to their area of conduction.

The auscultatory features of pericarditis form the subject of a short description, and a passing reference is made to aneurysm of the aorta.

103

THE HEART VALVES SEEN FROM BELOW.

The Outline, sketched from a dried specimen, represents a transverse section through the upper third of both ventricles. It aims at shewing the four valves in their mutual relations, as viewed from below—that is, from the ventricular aspect.

FIG. XXXIX.

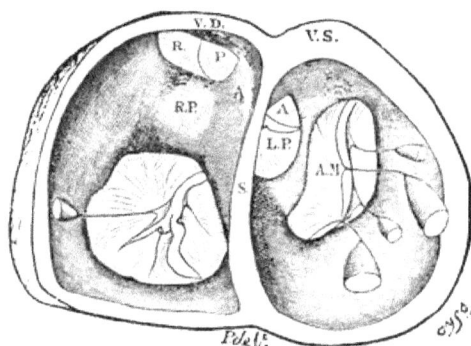

A.M. The anterior mitral flap, and next to it the posterior mitral flap.
A. The anterior aortic flap (in the right ventricle A shews the surface corresponding to this flap).
L.P. A portion of the left posterior aortic flap.
R.P. The surface corresponding, in the right ventricle, to the right posterior aortic flap, which is not itself in view.
P. A portion of the posterior pulmonary flap.
R. A portion of the right anterior pulmonary flap.
  The left anterior pulmonary flap is not in view.
  The tricuspid valve is fully displayed, but is not lettered.
V.D. The right ventricle in section.
V.S. The left ventricle.
S. The septum between the ventricles.

THE VALVES SEEN FROM ABOVE.

This is a view of the same specimen as in Fig. XXXIX., taken from above—that is, from the auricular aspect.

FIG. XL.

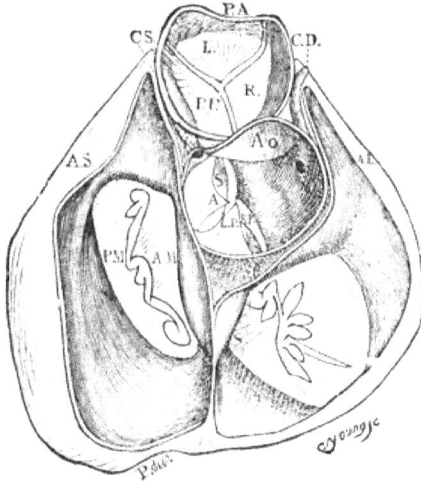

A.S.   The left auricular appendix in section.
A.I.   The right auricular appendix.
P.M.   The posterior mitral flap.
A.M.   The anterior mitral flap.
P.A.   The pulmonary artery.
L.   The left anterior pulmonary flap.
R.   The right anterior pulmonary flap.
P.P.   The posterior pulmonary flap.
Ao.   The aorta.
C.D. and C.S. The right and the left Coronary Arteries, with their ori-
         fices in the Sinuses of Valsalva.
A.   The anterior aortic flap.
L.P. and R.P. The left and the right posterior aortic flaps.
S., S., S., The interventricular septum distantly seen between the aortic
         flaps, and through the transparent membrane of the tricuspid
         valve.

THE SITES OF THE VALVES PROJECTED TO THE ANTERIOR SURFACE OF THE CHEST.

(The dermographic method does not attempt to indicate the exact situation of the several valves within the chest, but merely their projection. The grouping of lines which results is unreal—in the same way as the grouping of stars, such as it appears to the human eye, is unreal. The projection is nevertheless a help.

The valves appear as mere lines ; this also is a result of the projection.)

HOW TO "PLACE" THE VALVES IN THE PROJECTION. THE LEFT THIRD CARTILAGE AS A LANDMARK.

Outlines XIV. to XVII. shew how the projection of the semilunar valves can be localised by drawing a line (ML) from the middle point of the sternum to the left 2d chondrocostal junction. A more rapid guide, as accurate as the subject will admit (for individual varieties are common), is the left 3d chondrosternal junction. The horizontally arranged pulmonary semilunars follow the upper line of the cartilage, lying either a little way above or just below this line, and chiefly to the left of the sternum. The aortic semilunars cross the junction obliquely and lie chiefly under the left half of the width of the sternum. The projection of the mitral valve is also connected with the 3d cartilage. Beginning at its inferior border about ½ inch exterior to the sternum, it crosses the 3d interspace obliquely downwards and inwards and ends behind the sternum at the level of, and close to, the 4th chondrosternal junction.

The projection of the tricuspid is the only one not connected with the 3d left cartilage. It lies behind the sternum near its right border, in a nearly vertical line, extending from the level of the 4th to the upper level of the 6th cartilage.

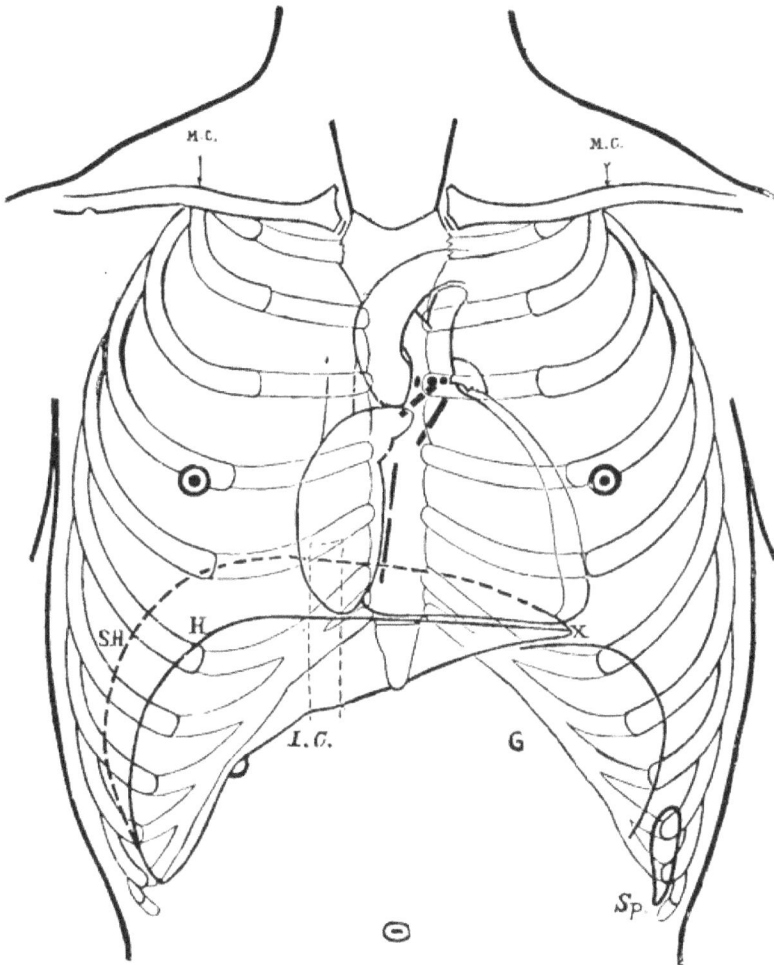

The three horizontally placed dots indicate the position of the pulmonary semilunar valve. The aortic valve forms an angle with the latter. The tricuspid valve is represented by the vertical line. The mitral valve by the oblique line lying across the third left interspace.

THE TOPOGRAPHY OF THE SEMILUNAR VALVES.

I. THE PULMONARY VALVE.

Leaving aside the surface markings, we shall now consider the relations of the valves within the chest.

(1) The Pulmonary Artery lies vertically behind the left 2d interspace ;

(2) It is covered only by a thin edge of lung, and by the serous membrane ;

(3) It is overlapped on either side by the tip of the auricular appendices ;

(4) Its segments lie obliquely across a line ML, uniting the appendices (the bi-appendix line).

(5) Its flaps are known as :　The right anterior flap, AP,
　　　　　　　　　　　　　　　The left anterior flap, AP,
　　　　　　　　　　　　　　　The posterior flap.

(6) The line of attachment of the posterior flap slightly overlaps (at an angle) the line of attachment of the anterior aortic flap.

II. THE AORTIC VALVE.

(1) The origin of the Aorta corresponds with the left 3d (inferior) chondrosternal angle, and with the adjoining surface of the sternum.

(2) Its inclination is shewn in the Outline ;

(3) It is separated from the sternum by the Pulmonary Artery, with which its anterior surface is intimately connected.

(4) The Aortic Valve lies almost exactly in the oblique line ML passing through the two auricular appendages.

(5) Its flaps are known as :　The anterior flap, AS,
　　　　　　　　　　　　　　　The right posterior flap,
　　　　　　　　　　　　　　　The left posterior flap.

(6) At its right corner the anterior flap, AS, crosses the line of attachment of the posterior pulmonary flap.[1]

---

[1] The two vessels being at this spot adherent to each other, it is almost impossible, in opening the heart, to cut through both of them without damaging either one or the other of the two flaps just mentioned.

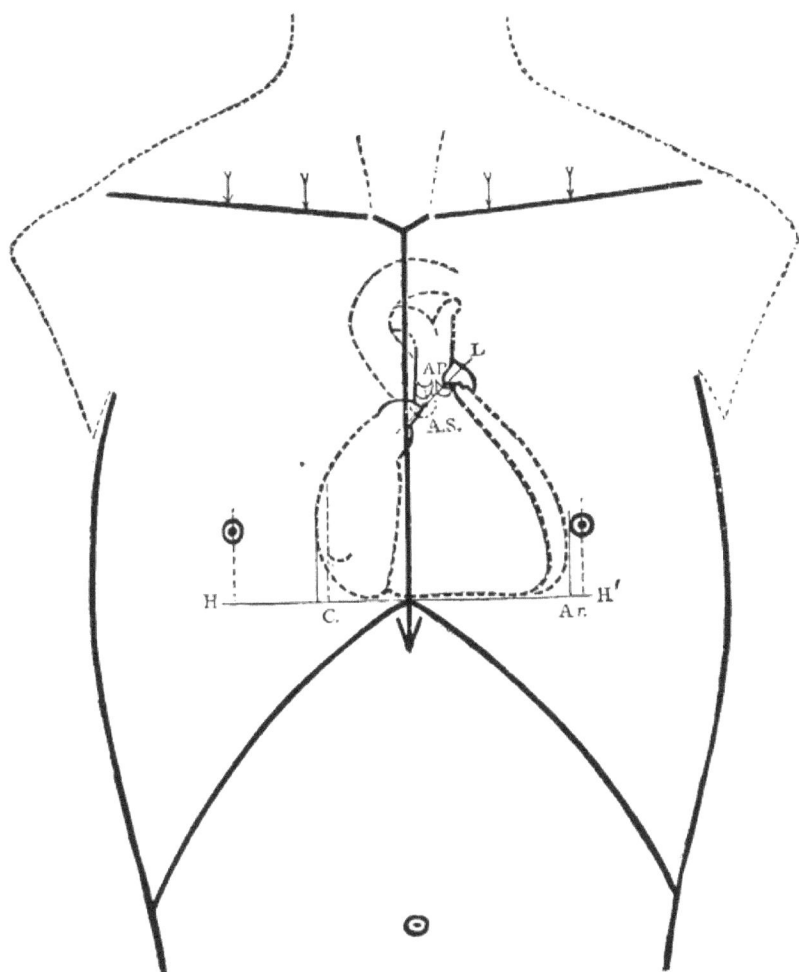

AP—The two anterior pulmonary valve flaps. AS—In dotted line the anterior aortic semilunar valve flap.

THE TOPOGRAPHY OF THE AURICULO-VEN-
TRICULAR VALVES.

I. THE TRICUSPID VALVE.

This valve is relatively superficial, since the right auriculo-ventricular groove lies immediately behind the front wall of the chest. The long diameter of the valve is almost vertical; but its upper extremity is rather more deeply placed than the lower.

Its right, left, and inferior flaps surround an oblong or oval orifice. The passage through, or the axis of, this orifice has an almost transverse direction, from left to right.

II. THE MITRAL VALVE.

This is the most deeply situated of all the heart-valves. It occupies the most remote corner of the left ventricle, posterior to and somewhat to the left of the aortic orifice.

Its long anterior flap is continuous with the fibrous aortic ring ; its posterior flap is attached to the posterior part of the left auriculo-ventricular ring. The axis of this orifice has a direction downwards, forwards, and to the left.

Its long diameter is not parallel with the anterior chest wall, and cannot be projected on to the latter without foreshortening. This explains why its projection should be so much shorter than that of the Tricuspid Valve.

FIG. XLIII.

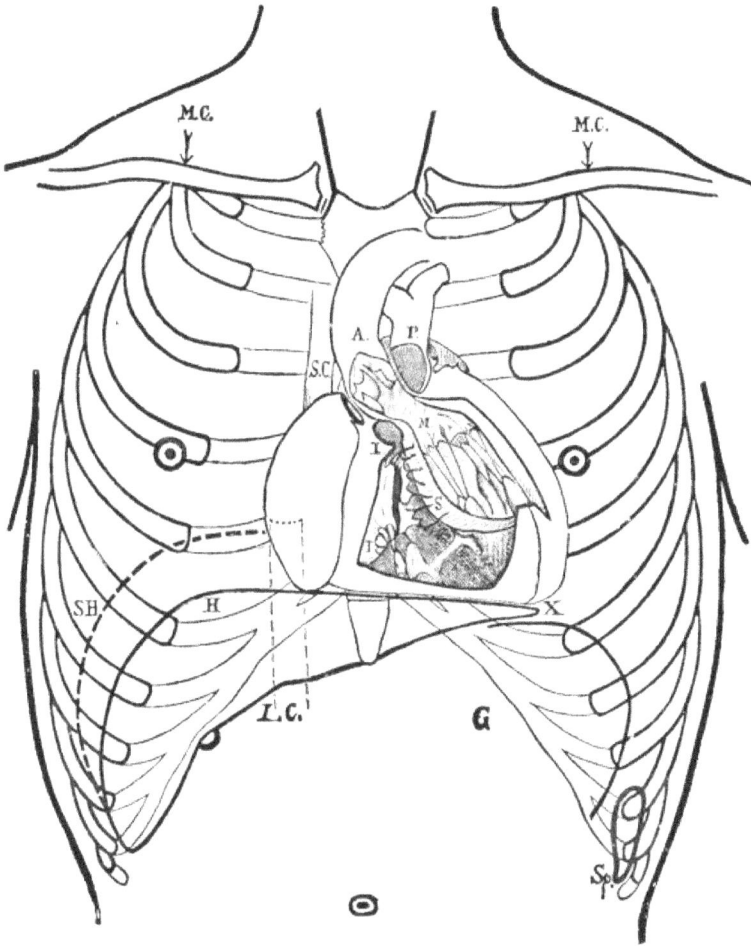

S—The septum cut down, affording a view into the left ventricle as well as into the right. M—The anterior mitral flap, behind which is seen a portion of the posterior flap. T—The inferior flap of the tricuspid ; the right and the left flap are also in view.

AUSCULTATION SERIES.

THE PRINCIPLES AND THE METHOD OF CARDIAC AUSCULTATION.

Cardiac Auscultation has three objects for study :

(1) The normal heart-sounds and their peculiarities.
(2) The pathological variations of the normal heart-sounds.
(3) The abnormal sounds known as " Bruits " or " Murmurs."

In other words, a clinical report of the auscultation of any heart implies answers to the following questions :

(A) Are the heart-sounds in any way peculiar without being abnormal ?
(B) Are the heart-sounds abnormal ?
(C) Are they accompanied by, or replaced by, any murmurs ?

It is part of the scheme of the graphic method to record these observations on a suitable thoracic Outline, by means of symbols. The symbols should be absolutely clear and unequivocal.

The beginner should be made to listen at first to absolutely normal heart-sounds, in subjects possessing a strong heart's action. At this stage he should practise "timing" the first and the second sound, by comparing their time with that of the apex-beat or of the carotid beat. Neglect of this essential detail of practical training accounts for the greater part of subsequent difficulties and delays.

The student's attention may be then called to the more striking varieties of the heart-sounds, such as the loud, the feeble, and the reduplicated sounds.

His next objects for study should be the simple valvular murmurs ; and murmurs should be selected for him well defined and easily heard. These he should practise " timing," that is, referring to the systolic or the diastolic period of the heart's cycle.

After the simple murmurs he should be made to study those of more complex rhythm. The timing of these is often a matter of difficulty even for trained observers. The following method is useful as a means of avoiding confusion when the heart's rhythm is misleading and, as it were, inverted :

Find the systolic time by carefully feeling the heart's impulse, or that of the carotid artery ; and beat time to this with the foot or with the finger. Auscultation of the doubtful sounds should then be performed whilst the observer continues beating time.

In some cases of loudly accentuated second sound, it may be easier to beat the diastolic instead of the systolic time, and to time the murmurs accordingly.

THE HEART-SOUNDS.

SUGGESTED CODE OF SYMBOLS FOR THE NORMAL HEART-SOUNDS AND FOR THEIR PATHOLOGICAL VARIATIONS.

(For the advanced student.)

(A) THE NORMAL HEART-SOUNDS.

The first Sound being relatively long has sometimes been figured thus ▬ and the second shorter Sound thus ◡.

The author has adopted this notation as the basis of an easy code of Symbols.

I.

Inasmuch as a first sound is produced both at the Mitral and at the Tricuspid Valves, and a second sound both at the Aortic and at the Pulmonary Valves, each of these may be conveniently described by coupling with the sign for the First Sound (▬), or with the sign for the Second Sound (◡), the initial letter of the Valve.

For instance, the Mitral First Sound would be expressed thus :

M

and the Aortic second sound thus :

A
◡

II.

Should the sounds happen to be weak or heard distantly, this may be readily expressed by substituting small characters and thinner lines for the capitals and for the thick lines ; for instance, a weak Mitral 1st Sound would be written thus :

m

Sounds so weak as to be almost inaudible might be expressed by dotted symbols.

FIG. XLIV.

SYMBOLS FOR NORMAL HEART-SOUNDS.

I

T M P A

II

t m p a

III

Ta Ma Tp Mp

IV

tA tP mA mP

V

TA TP MA MP

I—The four heart-sounds when heard loudly. II—The four heart-sounds when heard feebly (the dotted lines indicate very feeble sounds). III—The combination of a loud first sound with a feeble second sound. IV—The combination of a feeble first with a loud second sound. V—The first sound and the second both equally loud.

III.

If a loud Mitral 1st sound should be followed by a loud Aortic 2d sound two signs from the first series would be used. If one of the sounds should be weak and the other strong, this could be symbolised by combining a sign from the first with a sign from the second series, thus :

M a   and   **m A**

(B) INDIVIDUAL PECULIARITIES OF THE NORMAL HEART-SOUNDS.

Heart-sounds differ as do the voices of individuals—or their features. These differences can be described, or delineated, but it is quite impossible to express them by means of symbols. It would be premature for the junior student to attempt to study them. To the advanced student and to the physician they spontaneously become more and more obvious, with increasing experience and power of observation.

(C) PATHOLOGICAL VARIATIONS IN THE NORMAL HEART-SOUNDS.

These variations relate to :

    (1) Pitch and " timbre,"
    (2) Time or duration,
    (3) Loudness,
    (4) Uncoupling or reduplication,
    (5) Accentuation.

(1) Pitch and " timbre " must be described in words ; they are not easily symbolised.

(2) The same remark applies to time or duration, although, without introducing any fresh symbols, diagrams could be arranged displaying the normal series of the heart-sounds, and any series of abnormally long or abnormally short sounds, or lastly, of sounds separated by abnormally short or long intervals.

(3) As regards loudness, a suggestion has been made as to the mode of registering relative differences, such as occur in normal cases,

FIG. XLV.

SYMBOLS FOR REDUPLICATED HEART-SOUNDS.

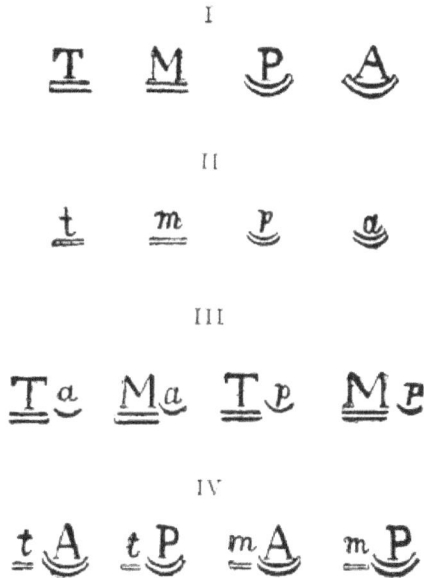

I

T M P A

II

t m p a

III

Ta Ma Tp Mp

IV

tA tP mA mP

I—Reduplication of sounds with loudness.   II—Reduplication without loudness.   III—Reduplication and loudness of first sound with feebleness of second sound.   IV—Reduplication of both sounds and loudness of the second sounds.

The sound may, however, be unusually loud, or feeble. A special notation is then required :

Excessive loudness can easily be represented by increasing the size of the capital letter, and the thickness of the line.

Extreme feebleness may be conveniently expressed by using a dotted or interrupted line, beneath a small letter (see Fig. XLIV.).

REDUPLICATION OF HEART-SOUNDS.

(4) The most common variation in time is relative delay or hurry of some constituent of the normal sounds. Each of the so-called heart-sounds being a blend of two sounds, delay or hurry of one of them leads to a breaking up of the composite sound into two,—or, as it is termed, to a reduplication of the sound.

Reduplications can be expressed by a very simple device—Let the line below the letter be repeated, as for instance in

$$\underset{\underset{\smile}{\smile}}{\mathbf{A}}$$

This symbol will signify, according to the code proposed, re-duplication of the Aortic Second Sound.

ACCENTUATION OF HEART-SOUNDS.

(5) Loudness and accentuation are most commonly associated, but they are not identical. A sound may be loud without that sharpness of delivery which we understand by accentuation. On the other hand, it is not uncommon for a heart-sound to lack loudness, whilst preserving a marked accent. The various combinations capable of arising and the mode of registering them are displayed in the accompanying illustration.

FIG. XLVI.

SYMBOLS FOR ACCENTUATED HEART-SOUNDS.

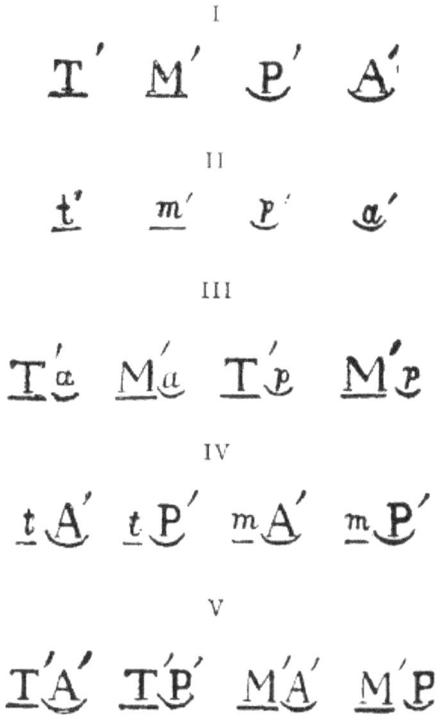

I

$$T' \quad M' \quad P' \quad A'$$

II

$$\underline{t'} \quad \underline{m'} \quad \underline{p'} \quad a'$$

III

$$\underline{T'\!\alpha} \quad \underline{M'\!\alpha} \quad \underline{T'\!p} \quad M'\!p$$

IV

$$\underline{t A'} \quad \underline{t P'} \quad \underline{m A'} \quad \underline{m P'}$$

V

$$\underline{T'\!A'} \quad \underline{T'\!P'} \quad \underline{M'\!A'} \quad \underline{M'\!P}$$

I—Accentuation and loudness of heart-sounds. II—Accentuation and feebleness of heart-sounds. III—Accentuation and loudness of the first sound, the second being feeble. IV—Accentuation and loudness of the second sound, the first being feeble. V—Accentuation and loudness of both sounds.

WHERE TO LISTEN FOR THE NORMAL HEART-SOUNDS

The normal heart-sounds may be heard over the whole præcordium ; not, however, with equal intensity. Each sound is heard most plainly over a definite area, and, as a rule, in the vicinity of the valve concerned in its production. The Outline shews the sites generally admitted to be the most favourable for auscultation of the normal sounds ; they will not require any description. The areas are more or less contiguous with the anterior projection of the valves : to this rule there is a striking exception, that of the Mitral valve. The Mitral, being deeply situated, is auscultated at the only spot, the heart's apex, where the left ventricle is superficial, although the distance from the valve is equal to the length of the ventricle. The aortic sound is also listened for at a slight distance from the aortic valve, in the right 2d interspace, which is for that reason known as the aortic interspace. The aortic valve is somewhat deeply situated behind the origin of the pulmonary artery.

FIG. XLVII.

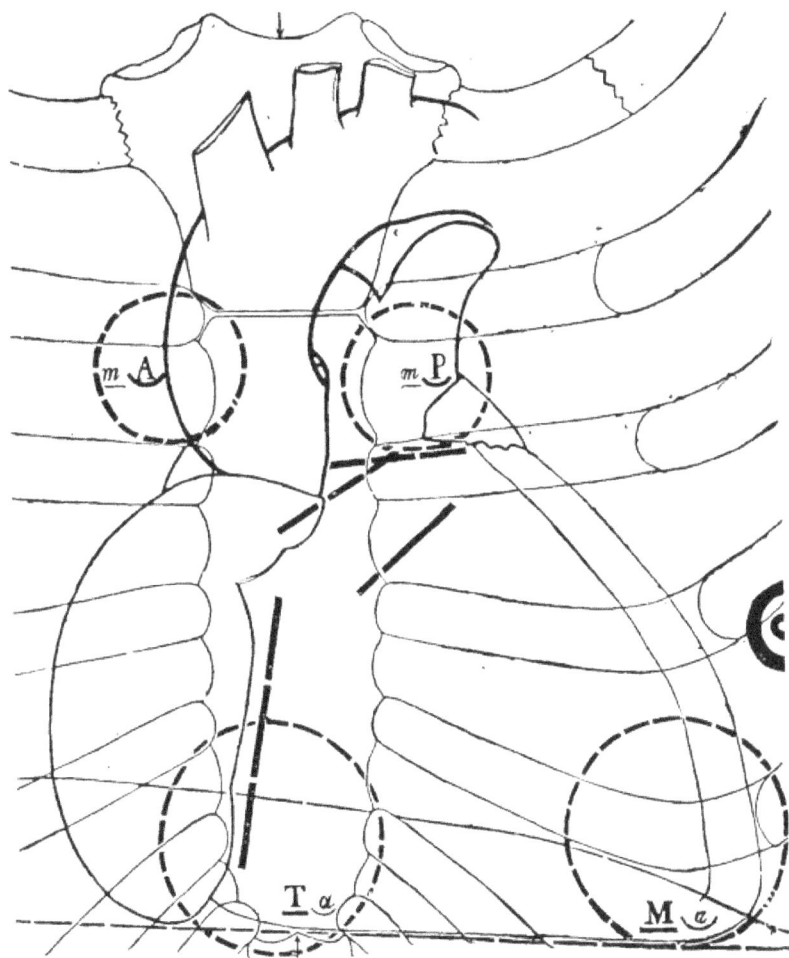

The sites of the valves projected to the anterior chest-wall are shewn as in a previous Outline.   Ma—The site of loudness of the mitral first sound. Ta—The site of loudness of the tricuspid first sound.   mA—The site of loudness of the aortic second sound.   mP—The site of loudness of the pulmonary second sound.

CARDIAC AUSCULTATION IN PATHOLOGICAL CONDITIONS.

THE GRAPHIC METHOD APPLIED TO THE STUDY OF CARDIAC MURMURS OR BRUITS.

The investigation of cardiac murmurs needs much time ; so does also the recording of results. The graphic method has the advantage of saving some of the time otherwise taken up in describing the examination ; it also tends to promote accuracy and completeness in examining the heart.

(1) Which is the valve affected ?
(2) Where is the murmur best heard ?
(3) Where does it cease to be heard ?

These questions must be answered in all cases where a murmur is audible ; and the answers are capable of being recorded graphically.

The nature and the site of intensity may be expressed by any code of symbols, provided the symbols are unequivocal. A code of this kind will be presently set forth.

A graphic representation of the *area of conduction* of murmurs generally takes place by means of shadings.

For quick work the simplest of all thoracic Outlines (Fig. I.) is the most useful ; but whenever accurate and searching observations have to be made, Outlines are required shewing the ribs and interspaces. It should be understood, however, that Outlines of this sort are not serviceable unless fairly correct anatomically, and of good size.

Most cardiac murmurs being valvular, it is important that the normal situation of the valves should be present to the mind of the auscultator. With that view in all the Outlines of the present series the valve sites have been displayed in addition to the murmurs.

FIG. XLVIII.

CODE OF SYMBOLS FOR THE NOTATION OF VALVULAR MURMURS.

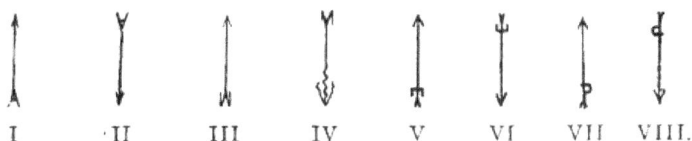

I—Onward aortic murmur. II—Regurgitant aortic murmur. III—Regurgitant mitral murmur. IV—Onward mitral murmur (with thrill limited to the apex). V—Regurgitant tricuspid murmur. VI—Onward tricuspid murmur. VII—Onward pulmonary murmur. VIII—Regurgitant pulmonary murmur.

The following modifications may with advantage be made in the old method of figuring bruits by arrows :

(1) The direction of the arrow must invariably indicate the direction of the blood-stream producing the bruit.

(2) The arrow should be provided with a letter denoting the valve implicated.

(3) If there be any thrill this may be symbolized by a waviness of the arrow.

(4) Loudness of the murmur may be expressed by increasing the thickness of the arrow ; softness of the murmur, by using a thin arrow or one with dotted line.

With these precautions several arrows may be used in the same Outline without fear of confusion, to denote the presence of different murmurs.

LIST OF THE CARDIAC VALVULAR MURMURS.

Valvular murmurs may be

(1) Organic (due to structural change in the valves or orifices) ; or

(2) Functional (due to faulty action of the muscles connected with the valves or orifices ; or to faulty quantity or quality of blood (hæmic murmurs).

THE FUNCTIONAL MURMURS.

The murmurs usually admitted to be functional are systolic in time, and aortic or pulmonary in site. They are generally considered to be hæmic in origin.

The pulmonary hæmic murmur is the most common of all murmurs.

THE ORGANIC MURMURS.

The way through any of the orifices may be constricted (Stenosis) ; or, a passage may be left by an inadequate valve at a time when the orifice should be closed (Reflux). Murmurs may therefore be produced,

By stenosis of :	By reflux through :
(1) The tricuspid orifice.	(5) The tricuspid orifice.
(2) The pulmonary orifice.	(6) The pulmonary orifice.
(3) The mitral orifice.	(7) The mitral orifice.
(4) The aortic orifice.	(8) The aortic orifice.

Of these eight varieties the first, the second, and the sixth are relatively infrequent. The others occur with considerable frequency and are described in the following Outlines.

It is important to realize that many of the murmurs commonly described as valvular, are really " orificial." This is especially the case with the functional group of valvular murmurs. Any valvular opening may become imperfect either through defect in the valve or through defect in its surroundings. A door for instance may fail to exclude draughts, either because damaged in itself, or if sound, because its framework has ceased to fit it. In the heart also an orifice may be at fault, and a murmur be set up without any defect in the valve.

THE SITES FOR THE AUSCULTATION OF THE CARDIAC MURMURS.

SYSTEMATIC EXAMINATION OF THE HEART FOR MURMURS.

Before a heart can be pronounced free from murmurs, either the entire præcordium must be auscultated inch by inch, or else its examination must be systematically localised to definite spots specially favourable for the recognition of murmurs. The position of the apex-beat is supposed to have been previously identified by palpation or by auscultation. The sites for auscultation are six in number and include, in addition to the areas which have already been assigned for the auscultation of the normal sounds, the *mid-sternum*, and lastly the 3*d* and 4*th left interspaces in the parasternal line* which it is convenient to term collectively the " Ventricular Site," this area being the centre of the anterior projection of the ventricular portion of the heart. Thus the stethoscope will have to be applied in succession to :

(1) The apex for mitral murmurs,

(2) The right 2d interspace close to the sternum for aortic murmurs,

(3) The left 2d interspace for pulmonary murmurs, and for Naunyn and Balfour's murmur,

(4) The lower end of the sternum for tricuspid murmurs, and also for aortic regurgitant murmurs.

THE SITES FOR AUSCULTATION—*Continued.*

(5) The mid-sternum for tricuspid, for pulmonary, and especially for
    aortic regurgitant murmurs,

(6) The "ventricular site" for aortic, for mitral, and for hæmic
    murmurs.

No endocardial murmurs will escape this search, and it is improbable
that any exocardial murmurs could pass unobserved.

THE AREAS OF CONDUCTION OF MURMURS.

The following Outlines are devoted to a study of the areas of conduction
of the cardiac murmurs. The mode of conduction is characteristic in each
case ; in some of the murmurs it is so characteristic that a diagnosis of the
variety of murmur present may rest upon this feature alone.

FIG. XLIX.

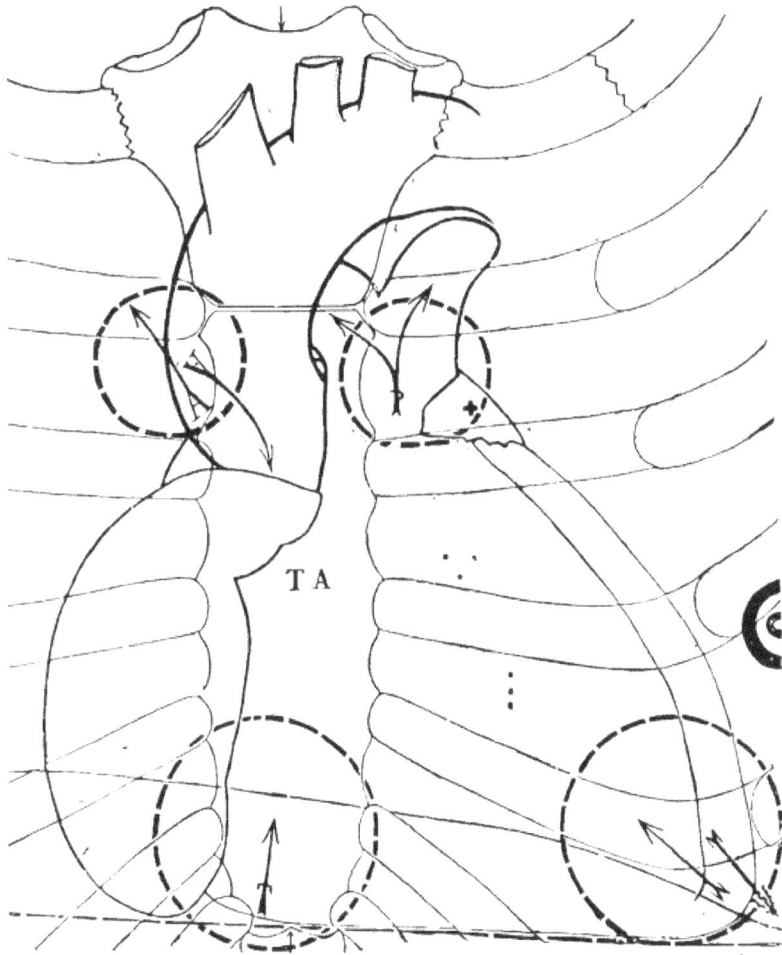

The apex-circle—seat of the mitral murmurs. The lower sternal circle—seat of the tricuspid regurgitant murmur. The circle at the right interspace—centre for aortic murmurs. The circle at the left interspace —centre for pulmonary murmurs and for Naunyn and Balfour's murmur. TA—at the mid sternum, is a common seat for the regurgitant aortic murmur. ˙.˙ and ⁝ form part of the "ventricular site," where the regurgitant aortic murmur is sometimes heard, and mitral and hæmic murmurs may also be audible.

THE SITE AND AREA OF CONDUCTION OF THE SYSTOLIC PULMONARY MURMUR, AND OF THE HÆMIC MURMUR.

Putting aside cases of congenital disease, organic murmurs of the pulmonary orifice are uncommon ; especially that of regurgitation, which has not been included in the Outlines.

The characters of a Systolic (onward or obstructive) Pulmonary Murmur are :

Intensity at the site of the Pulmonary Valve,

Extension along the left 2d and 3d cartilages, and the 2d and 3d interspaces, and towards the right as far as the right edge of the sternum.

This description applies also to the hæmic murmur which is often a very "rough" sounding one. Its mechanism is still doubtful, but most authorities agree in localising the production of sound in the Pulmonary Artery. This is not the view taken by the Naunyn and Balfour.

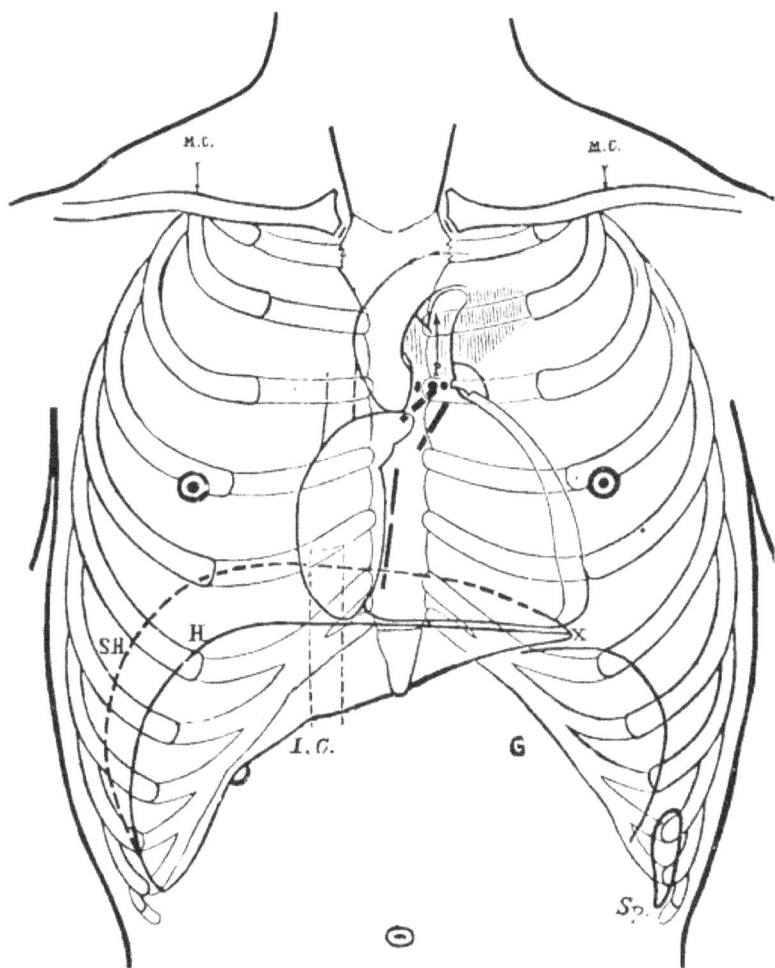

The arrow represents the pulmonary systolic murmur, and the shading its usual area of conduction.

OUTLINE RELATING TO NAUNYN AND BAL-FOUR'S MURMUR.

Naunyn and subsequently Balfour failed to obtain proof of the localisation just mentioned. They propounded the view that a functional murmur was apt to be produced by a dilatation of the mitral orifice, forming part of a general dilatation of the heart, and allowing backward injection of the already full Left Auricle.

The cross placed in the Outline of the Left Auricular appendix is intended to remind the student of this important theory, a discussion of which does not belong to the scope of these pages.

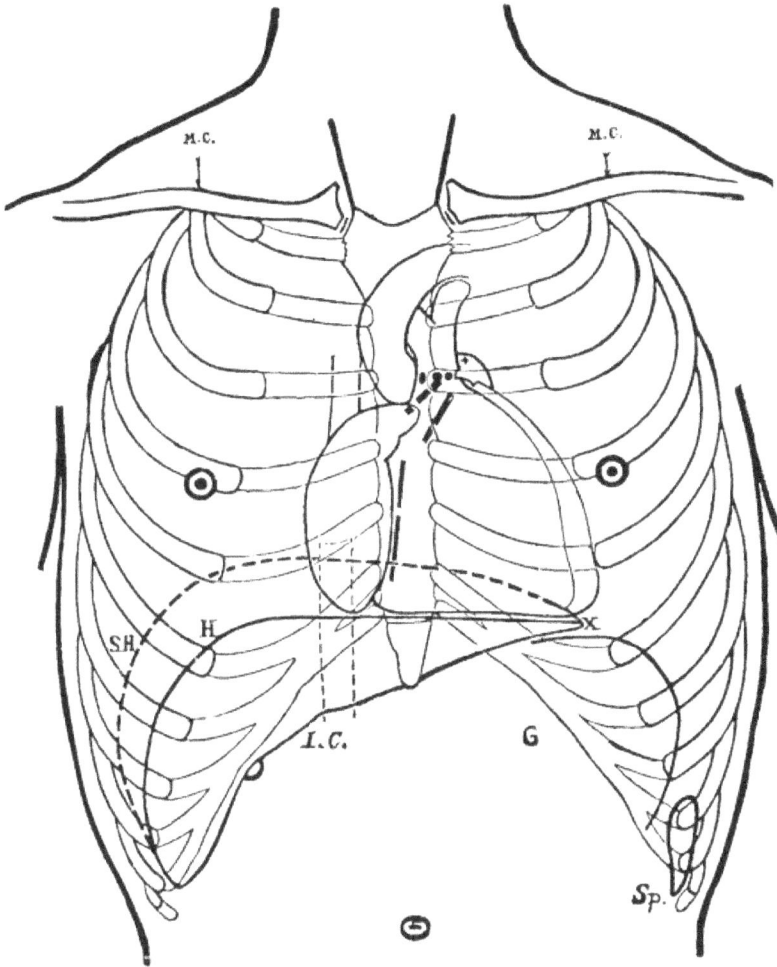

The small cross draws attention to the left auricular appendix, which is supposed to be the seat of regurgitation from the left ventricle.

THE SITE AND AREA OF CONDUCTION OF THE TRICUSPID REGURGITANT MURMUR.

Tricuspid incompetence is extremely common; loudness of the tricuspid regurgitant murmur, much less common.   A knowledge of this fact should keep us on the alert lest we should pass over a murmur which was only just audible.

The conditions which favour the wide conduction of the mitral systolic murmur fail in this case ; and the overlapping of the right over the left ventricle is a likely source of confusion between the two murmurs and a special reason for care in diagnosis.

WHERE TO LISTEN FOR THIS MURMUR.

As indicated by the shading, the base of the xiphoid and the lower extremity of the sternum, especially on the right side of the middle line, are the sites of intensity.   This is the place for *auscultation*, although the murmur may be conducted elsewhere ; for a mitral murmur is probably never conducted to this spot, whilst a tricuspid murmur is always loudest here.

The arrow in the Outline is not placed in the direction of the regurgitant blood stream, but rather in the direction of the sternal conduction of the murmur.

The characters of a Tricuspid Regurgitant Murmur are :

(1) Occurrence at the moment of the Ventricular Systole ; (thereby the 1st sound may or may not be completely replaced.)

(2) Blowing character (bellows sound) ;

(3) Intensity greatest over the xiphoid cartilage, and tip of the sternum ;

(4) The bruit is very superficial ("near the ear") ;

(5) It is conducted for some distance upwards along the sternum, and sometimes to the left of that bone.

N. B.   A Direct (onward) Tricuspid Murmur is very uncommon.

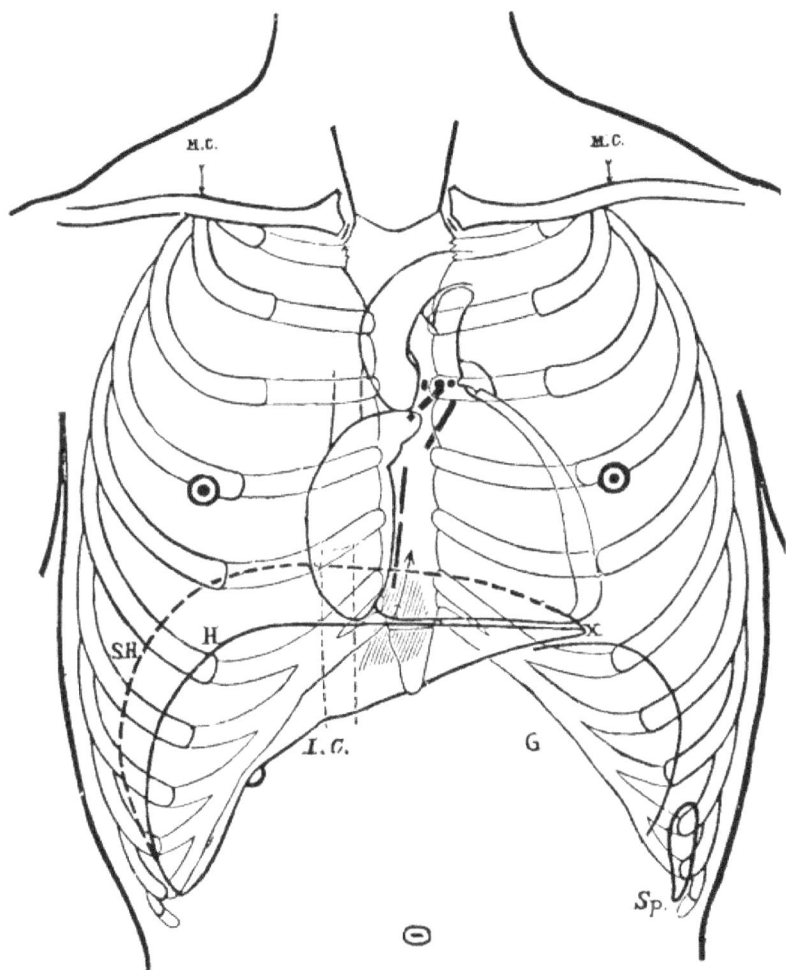

The arrow represents the murmur, although not the direction of the regurgitant blood stream. The shading gives the area of conduction, which may, however, be more considerable than here shewn.

THE SITE AND AREA OF CONDUCTION OF THE ONWARD AORTIC MURMUR.

The site of intensity extends in this case above the level of the valvular orifice, which is rather deeply situated behind the 3d left sternal junction and adjoining portion of the sternum. The murmur is conducted loudly over the shaded surface, owing to the strength of the ventricular contraction, and owing to the close and extensive relation of the aortic arch and of its branches to the sternum and to the upper costal cartilages. Often the murmur is not only locally a very loud one, but it is audible over a very large surface, or even over the whole chest. A thrill is a frequent accompaniment of an extensive murmur of this kind.

In the Outline the shading has been restricted to the area which is usually described as that of conduction for aortic systolic murmurs, and over which the latter are always audible, if audible at the site of intensity.

It is an important diagnostic feature of this murmur that it is loudly conveyed along the carotids. The same is true of a hæmic aortic murmur, but the general clinical features of the case, as well as the cardiac symptoms, seldom allow any confusion between the organic and the functional bruits.

The peculiarities of sound special to the aortic systolic murmur vary ; it may be harsh, blowing, noisy, sometimes musical and sometimes accompanied with grating or knocking sounds.

In conclusion the characters of the Onward Aortic Murmur are :

(1) Occurrence at the moment of systole (with or without complete replacement of the 1st sound) ;

(2) Relatively long duration ;

(3) Loudness ;

(4) Superficial type ;

(5) Variously harsh, rumbling, grating, or musical character ;

(6) Intensity greatest over the 2d right cartilage and interspace ;   .

(7) Conduction chiefly upwards to the root of the neck ; (but it may extend over the left upper pectoral region, as well as over the right, or even over the whole front of the chest.)

(8) Conduction along the carotid arteries

(9) Very commonly a thrill.

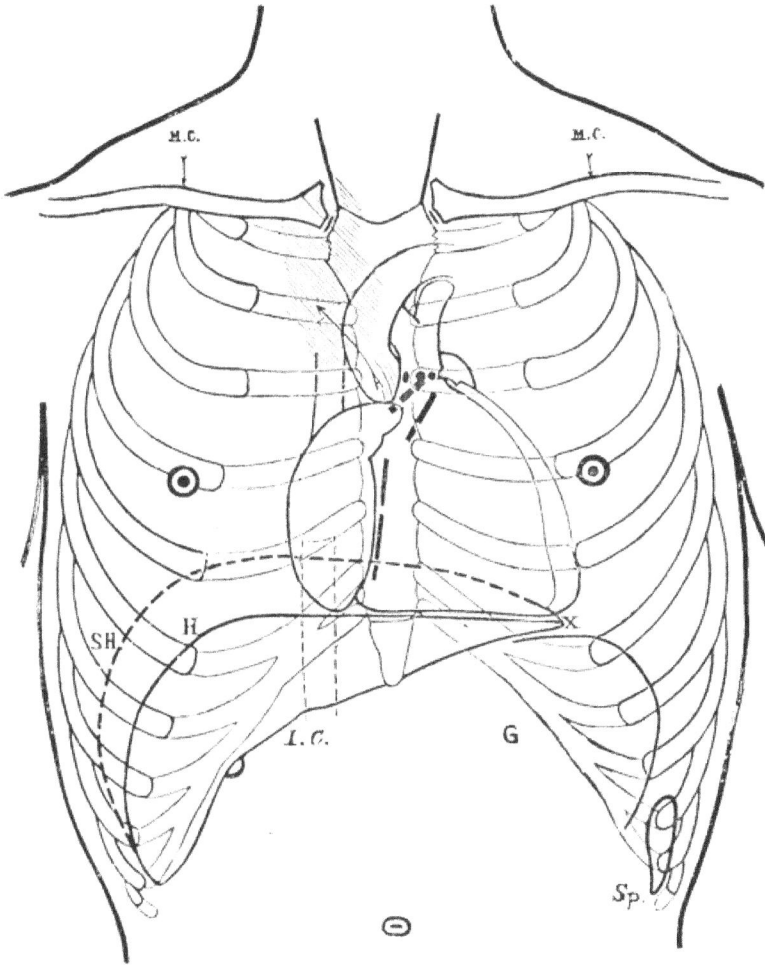

FIG. LIII.

The arrow indicates the direction of the blood stream producing the onward aortic murmur. The shading corresponds with the minimum area of conduction.

THE SITE AND AREA OF CONDUCTION OF THE REGURGITANT AORTIC MURMUR.

This bruit resembles a breath blown quickly from the throat without approximating the lips. It has been compared to the noise of an escape of steam. To put the matter more precisely its peculiarities are:

> Evenness,
> High pitch,
> Softness, or breath-like nature,
> Absence of sonorous vibrations,

a combination not presented by any other murmur.

The softness and slight intensity of this bruit call for special attention in searching for it. It is of all murmurs the most apt to be missed owing to a faulty or unsuitable stethoscope: a caution to be laid to heart, not only by the student undergoing examinations, but by the practitioner in his more responsible duties.

WHERE TO LISTEN FOR THIS MURMUR.

Its site of intensity is below the level of the aortic valves, over the mid-sternum, and over the sternal end of the right 4th cartilage. It is not commonly louder beyond the sternum than over that bone, although this sometimes occurs.

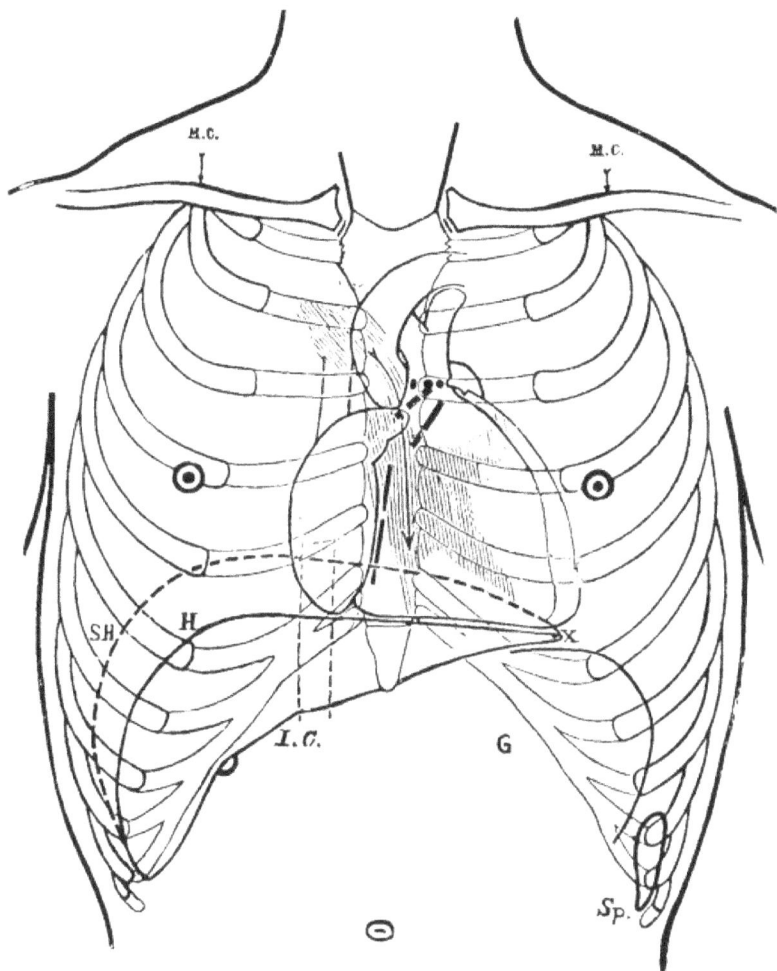

The direction of the arrow is that of the blood reflux producing the more common form of diastolic aortic murmur. The shading indicates the three directions in which the murmur may be conducted.

Its conduction may be over a wide area, as shewn in the Outline, but more often it occupies a portion only of the shaded surface.

Upwards :   It may extend as far as the 1st right interspace.

Outwards :   Conduction may take place horizontally towards the left.

Downwards :   The murmur may extend as far as the apex.

It has been described as occasionally heard at the apex only. This is the exception. The rule is for the murmur to be most plainly heard over the lower half of the sternum. Indeed it is practically as loud over the lower sternal third as higher ; and this situation may be regarded as the site of election for listening for the murmur, being alike remote from the area of the respiratory and from that of the cardiac sounds. Whilst being conducted by the sternum, the sound is sometimes perceptibly modified by that bone.

The diversity in the mode of conduction of aortic regurgitant murmurs is ascribed to the fact that the lesion may be limited to any one of the three segments, or to two of them, or may extend to all three.

Sir Walter B. Foster long since explained on these lines the occasional conduction to the apex of a regurgitant murmur, the right posterior segment being then supposed to be at fault.

N. B. The danger of mistaking for an aortic murmur a diastolic murmur produced at the mitral orifice should be borne in mind.

The distinctive features of an Aortic Regurgitant Murmur are :

(1) Occurrence at the moment of the normal closure of the Semilunar valves (the aortic second sound may or may not be completely replaced) ;

(2) Continuance of the bruit during the first part of the diastolic silence, with gradual subsidence ;

(3) Breath-like character, of various but usually rather high pitch ;

(4) Soft and distant character ;

(5) Intensity greatest behind the sternum, at the level of the second inter-space ;

(6) Conduction chiefly down the sternum ;

(7) Occasional conduction towards the left nipple, or towards the apex.

Roughness, loudness, musical character, and thrill are rarely present.

THE SITE AND AREA OF CONDUCTION OF THE REGURGITANT MITRAL MURMUR.

This is the most common of all organic valvular murmurs.

Its loudness and its quality vary greatly according to the extent and more particularly to the variety of the lesion. It is not breath-like as the aortic regurgitant murmur, but blowing (bellows sound). It may be soft blowing, hard blowing, rough, or even musical.

Its diagnosis from exocardial and from hæmic murmurs is by no means always easy.

WHERE TO LISTEN FOR THIS MURMUR.

Always at the apex. But, inasmuch as other systolic murmurs are also audible here, careful discrimination is needed.

The site of intensity of the mitral systolic murmur is the apex : this is not the case with the aortic systolic murmur, nor with the tricuspid, nor, as a rule, with the hæmic murmurs.

The conduction as shewn in the shaded part of the Outline is outwards and slightly upwards towards the axilla.

The arrow shews somewhat roughly the direction of the regurgitant blood stream. The murmur may also be heard at the angle of the left scapula and to the left of the 6th dorsal vertebra. The reason for this is obvious when we look at a lateral view of the heart.

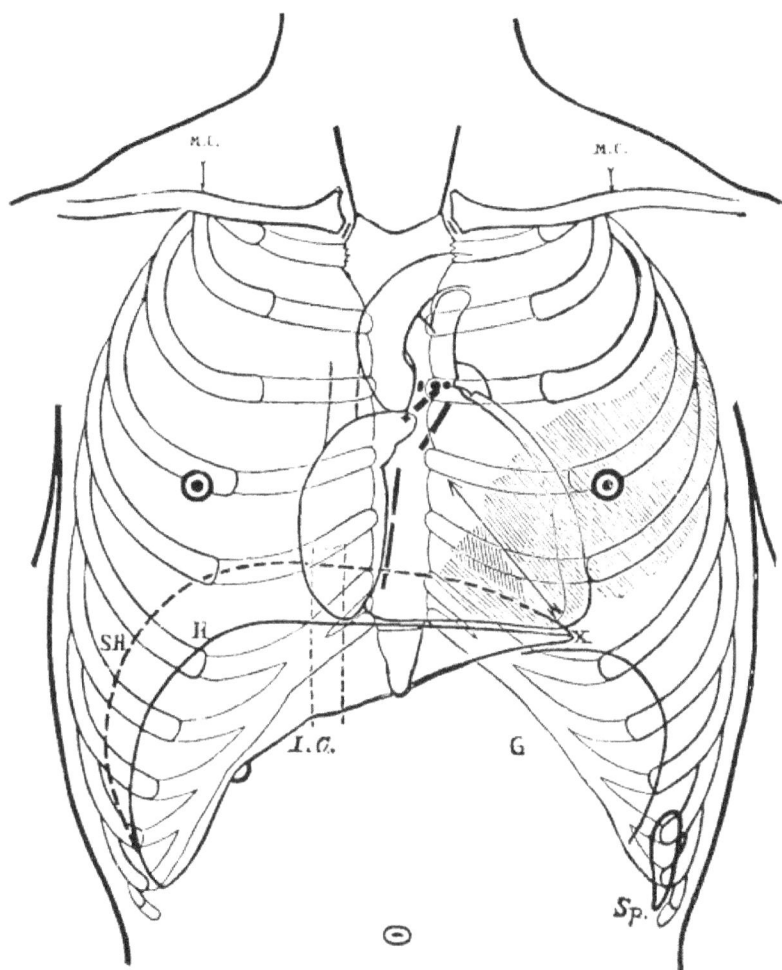

The arrow shews the direction of the blood reflux. The shading indicates the area of conduction of the murmur.

THE HEART VIEWED FROM THE LEFT SIDE TO ILLUSTRATE THE MODE OF CONDUCTION OF THE REGURGITANT MITRAL MURMUR.

The Outline is intended to shew the extensive surface of the left ventricle which presents in the left half of the chest, and transmits the murmur outwards to the lateral wall of the thorax. The auricle into which regurgitation occurs is doubtless the channel for the conduction of the murmur backwards. The size of the heart in this Outline is disproportionate, but it will be remembered that hypertrophy is among the results of mitral incompetence.

Conduction towards the right also takes place, but does not, as a rule, extend much beyond the parasternal line ; although owing to its loudness a mitral murmur may not infrequently be distantly audible beyond the latter.

Tricuspid regurgitation being commonly associated with advanced *mitral incompetence, both murmurs may be blended* at the left parasternal region. Their simultaneous existence will be manifest if the systolic murmur gains in intensity towards the right chondro-xiphoid angle, whilst on the other hand it attains a maximum at the apex of the heart.

A purely hæmic bruit will be audible over the left upper portion of the præcordium, but almost invariably it will be loudest at the 2d left interspace close to the sternum.

Lastly an aortic systolic murmur, which is often propagated to the whole præcordium and even beyond its limits will be much more intense over the right 2d interspace and cartilage, and will be conducted into the carotids (*cf.* p. 134).

FIG. LVI.

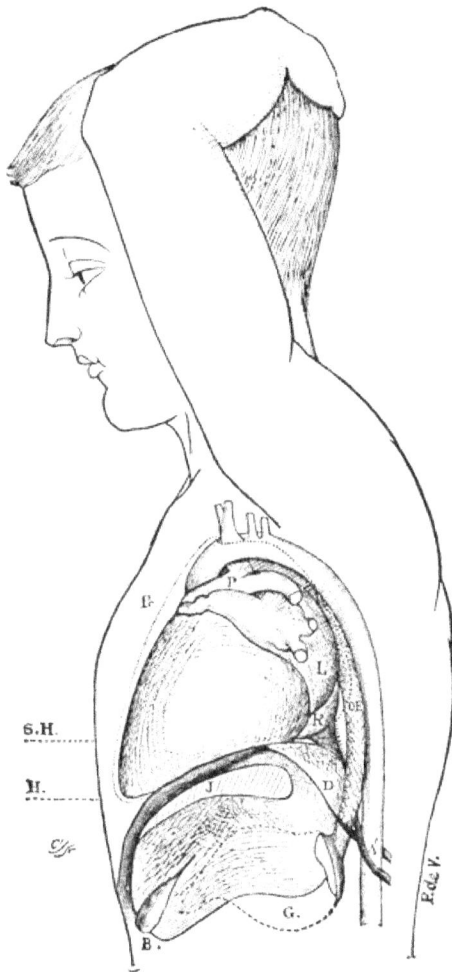

Dilated and hypertrophied heart in a child. The greater part of the convex surface of the left ventricle faces the axillary region. It would the more closely approach the latter the greater the degree of the cardiac hypertrophy. The left aspect of the left auricle is also in view, but the bulk of the auricle faces the spine.

THE SITE AND AREA OF CONDUCTION OF THE MITRAL AURICULAR-SYSTOLIC, PRÆ-SYSTOLIC, OR DIASTOLIC MURMUR.

This murmur may occur alone ; more commonly it is complicated with a mitral systolic murmur.

I.

UNCOMPLICATED AURICULAR-SYSTOLIC MURMUR.

The Outline illustrates the downward and outward direction of the blood stream in the left ventricle and the thrill to which it gives rise at the apex of the left ventricle.   The circle denotes the strict limitation of the murmur to the apex region.

The following are among the chief characters of this murmur :

Inconstancy in occurrence ;
Variability in quality ;
Diversity of sounds audible in different cases (see below) ;
Limitation to a small region, at most 3″ in diameter, including the apex ;
Accompanying thrill ;
Forcible systolic impulse of the apex ;
Abrupt loudness of the systolic sound at the apex ;
Loudness of the 2d pulmonary sound ;
Absence of conduction towards the axilla and to the back ;
Displacement of the apex outwards and slightly downwards (as indicated
    by the arrow).

WHERE TO LISTEN FOR THIS MURMUR.

Auscultation should be made at the apex, and preferably a little external to the usual apex site, for it will be found that the apex beat is not limited to a single point but presents a rather considerable surface.

FIG. LVII.

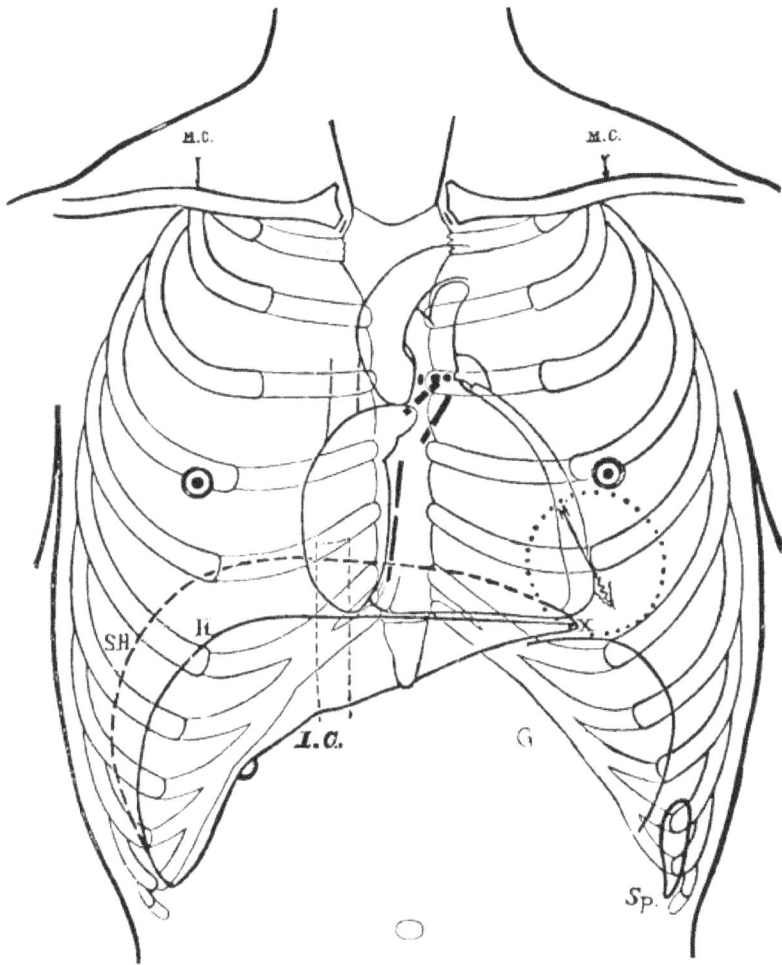

The arrow indicates the direction of the blood stream during the period of murmur. The limited area over which the murmur extends is circumscribed by the circle. The localised thrill is expressed by the waviness of the head of the arrow.

The Three chief Varieties of auricular-systolic murmurs are :

(1) The early diastolic murmur,

(2) The mid-diastolic murmur,

(3) The late diastolic, or præsystolic murmur in the strict sense.

Additional varieties arise according as murmurs (1) and (2) are, or are not, continued through the diastolic interval.

A discussion of the causes of these several varieties would be at the best speculative, and, for that reason alone, ill suited for the scope of this book.

DIFFERENTIAL DIAGNOSIS OF AURICULAR-SYSTOLIC MURMURS.

The least equivocal variety is the mid-diastolic, when the murmur is conducted up to a loud 1st sound and followed by a loud pulmonary 2d sound. Confusion is most liable to occur in the case of the early and in the case of the late diastolic (or præsystolic) murmur. The first of these might be mistaken for an aortic regurgitant murmur, especially if (as recorded in rare cases) it should be audible at the apex only.

The late murmur might be so closely followed by the systolic event as to suggest that it formed part of the latter. In many cases of this kind individual opinion based upon individual perception for sound and time is the *ultima ratio.*

In both cases assistance will be derived from the presence or absence of the other physical signs of mitral stenosis (described above) ; and from the presence or absence of those belonging respectively to mitral or to aortic incompetence. The pulse, which is distinctive in all three conditions, is a valuable help.

And lastly, the symptoms may tell strongly in one or the other direction. It should not be overlooked :

(1) that all three varieties may be imitated by exocardial sounds ;

(2) that the præsystolic murmur may be closely imitated by the heaving and delayed first sound of an hypertrophied heart confined by adhesions ;

(3) that the auricular-systolic murmur may be "cogged" or "interrupted" and may thus be thought to be a reduplicated second sound.

II.

AURICULAR-SYSTOLIC MURMURS ASSOCIATED WITH OTHER MURMURS.

Very commonly a mitral diastolic murmur is associated with another, or even with two other murmurs. The ordinary complication is that of a mitral systolic murmur; and this may deprive us of a helpful sign for diagnosis, viz., the snapping or knocking First Sound. An inevitable complication eventually follows, viz., backflow through a dilated tricuspid; unless the latter, as sometimes occurs, should have become secondarily stenosed. The dilated tricuspid orifice will give rise to a murmur; but tricuspid stenosis will seldom be sufficiently complete to give rise to an audible bruit.

Aortic valvular disease is a complication by no means rare.

Pericardial friction has already been mentioned.

Lastly, when cardiac failure has supervened the cardiac rhythm may audibly suffer from the strain, and loud reduplication of one of the sounds is the result. The confusing whirl of cardiac sounds thus produced has been termed by the French "bruit de galop": the heart's rate is usually very high under these circumstances, and all distinctive murmurs may for a time be submerged.

Short of the climax of confusion just described any superadded murmur will be a serious obstacle to strict definition of this bruit by auscultation. There is, however, in stenosis a peculiarity of the heart's action as a whole, which seldom misleads the experienced ear, though individual sounds may be too rapid for due appreciation.

THE AUSCULTATORY SIGNS OF CARDIAC HYPERTROPHY.

No bruits are connected with hypertrophy pure and simple. At most the normal sounds are more or less modified (see p. 116). The first sound is exaggerated, ponderous, prolonged, and heaving ; the second sound, more or less accentuated. A reduplication of the first sound, heard near the apex, has been pointed out by Dr. Broadbent as a sign of albuminuria, and will often be found in cases of renal hypertrophy of the left ventricle.

The cardiac rhythm, if altered, is slowed rather than accelerated. None of these signs, however, belong to hypertrophy exclusively. Palpation and percussion must be called to aid for its diagnosis, which is often also much assisted by the general symptoms.

THE AUSCULTATORY SIGNS OF CARDIAC DILATATION.

There are no bruits distinctive of cardiac dilatation *per se*, although symptoms arising from the associated valvular incompetence are usually present.

On the other hand, the normal action and sounds are decidedly altered.

(1) The rhythm is accelerated and may be irregular ;

(2) The ventricular impulse, although unduly prominent and (except in cases where dilatation is due to pre-existing pulmonary emphysema) very extensive, lacks the hard feel of the hypertrophied ventricle ;

(3) The 1st heart-sound is feeble, short, and sometimes faltering. The shortness of the 1st sound arises from the spasmodic effort of the overtaxed ventricle, which endeavours in vain to empty itself completely ;

(4) The weakness of the second sound is connected with the low pressure under which the aorta is injected ;

(5) The overcharged ventricle hastens to contract again : the long pause is thereby shortened ;

(6) A frequent result is an apparent equality of the 1st and of the 2d sound as regards strength and quality, and of the 1st and 2d pause as regards duration ;

This resemblance with the fœtal heart's action (embryocardia) is strengthened by the rapidity of rate which the dilated heart has in common with the fœtal.

(This description applies to cases of considerable dilatation affecting the whole heart ; it would have to be modified in cases where the dilatation was unilateral or complicated with hypertrophy.)

AORTIC ANEURYSM.

THE STRUCTURES LIABLE TO IMPLICATION IN ANEURYSM OF THE ASCENDING AORTA AND OF THE ARCH.

Aortic Aneurysm is a very large subject, rendered all the more difficult to treat owing to the diversity of individual cases. A systematic account of the whole is not therefore attempted. The accompanying Outline has regard only to the important anatomical questions which have to be considered in all cases of aneurysm of the arch or of the ascending aorta. It has for its purpose to help the student to realise the mechanical results of the growth of tumours of this sort.

Leaving aside the intracardiac aneurysms and aneurysms situated just above the valves, the pressure from which takes effect upon the substance or the cavities of the heart, we shall consider only those whose growth is well above the sinuses of Valsalva.

A feature of all aneurysms of this class is that the aortic 2d sound is "thudding," *i.e.*, very loud and very much accentuated. The large symbol **A** has reference to this fact.

The Vena Cava Superior, although a most important structure in relation to aneurysm of the first part of the aorta, could not be introduced into the diagram ; it should have occupied the site of the symbol **A.**

PRESSURE EFFECTS.

In aneurysm of the 1st portion pressure is apt to bear :

   (1) forwards—on the anterior chest-wall, etc.

   (2) to the right—on the Superior Vena Cava, and on the lung.

   (3) backwards—on the right bronchus, or trachea ; and on the right pulmonary artery.

In aneurysm of the Arch pressure tells :

   (1) downwards and backwards—on the left bronchus, or on the trachea ; on the left, or on the right pulmonary artery ; and on the left recurrent laryngeal nerve.

   (2) forwards and upwards—on the left Innominate Vein.

   (3) to the left—on the upper part of the left lung.

Dyspnœa may thus result either from direct compression and finally ulceration of the left bronchus or trachea, or indirectly owing to stretching of the left recurrent laryngeal nerve.

FIG. LVIII.

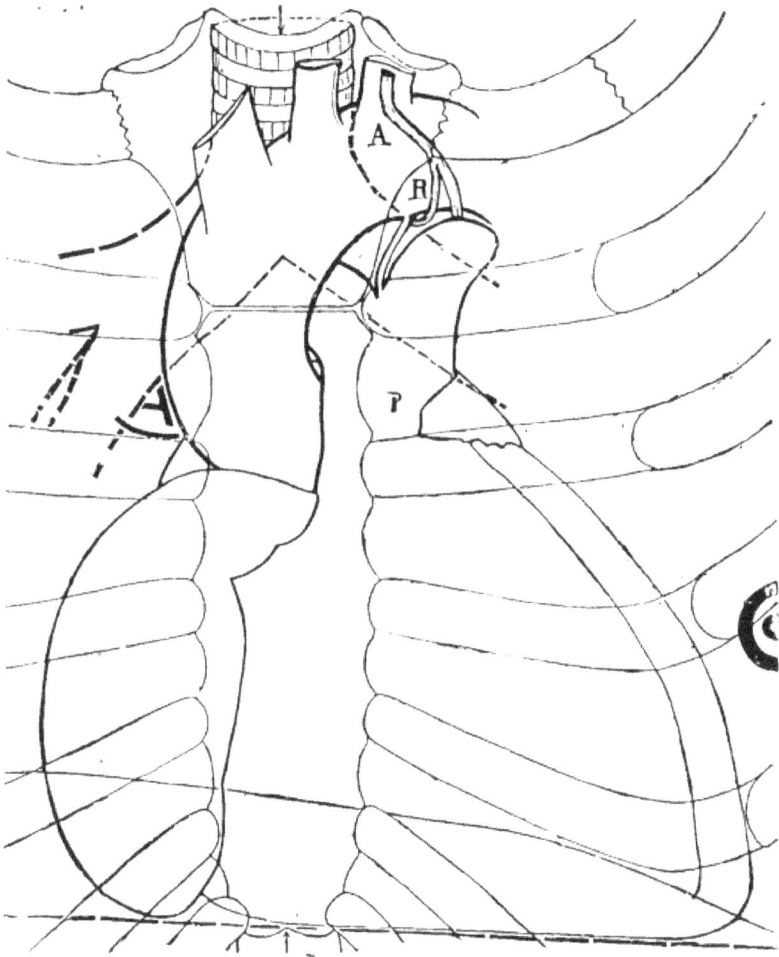

A—Arch of the aorta. P—Pulmonary artery. R—Left recurrent laryngeal nerve, in contact with the aortic end of the ductus arteriosus. A—Site of loud thudding aortic second sound.

N. B.— *The superior vena cava should have been shown along the right side of the ascending aorta. The trachea and main bronchi, and the other structures, will be easily recognised. The arrows indicate the middle line.*

PERICARDITIS.

A BRIEF SUMMARY OF ITS VARIETIES AND TYPICAL SIGNS.

I.

LOCAL OR LIMITED PERICARDITIS.

The affection is generally not severe ; commonly it remains latent and is not diagnosed. It results in :

Adhesion, or
Roughening of surfaces.

LIMITED PERICARDIAL ADHESIONS.

These are often not capable of recognition during life.
Clinically they are not regarded as of serious import.

LIMITED PERICARDIAL ROUGHNESS.

(1) This may be known by the occurrence of friction sounds over a limited area.

(2) During its earlier stage it may give rise at first to a softer variety of sound, the " pericardial (or exocardial) murmur."

(3) The roughness ultimately disappears, leaving only a smooth, thickened, or opaque patch on the pericardial membrane ; this is one form of the " white-patch " or " milk-spot."

II.

GENERAL PERICARDITIS.

(A) SIMPLE OR UNCOMPLICATED GENERAL PERICARDITIS.

Of this there are two varieties :

Dry or fibrinous pericarditis,
Pericarditis with effusion.

(B) GENERAL PERICARDITIS OF COMPLICATED OR OF SPECIFIC TYPE.

Pericarditis may be the result of specific causes such as :

> Pyæmia,
> Tuberculosis,
> Carcinoma, etc.

These forms of the disease may or may not follow a course, and give rise to signs analogous to those of the simple variety.

DRY OR FIBRINOUS PERICARDITIS.

This may be, as previously stated, only a stage in pericarditis ; or it may constitute the whole disease. The fibrin thrown out at the surface of the membrane leads to one of two results :

> (a) Pericardial agglutination or adhesion ; or
> (b) Pericardial friction and attrition.

(a) PERICARDIAL AGGLUTINATION OR ADHESION.

Adhesions may arise :

(1) Between the heart and its sac (*intra-pericardial adhesions*) ; or
(2) Between the pleuro-pericardium, or lateral layer of the sac, on the one hand, and the chest-wall or the median surface of the lung on the other (*extra-pericardial adhesions*) ; or, lastly,
(3) Between all these structures simultaneously, the heart becoming adherent to its sac, and the sac to its surroundings (*intra-* and *extra-pericardial adhesions*).

Pericardial adhesions vary in their consistence and in their length, and therefore in their effect upon the heart. Pericardial agglutination implies an absolute fusion between the opposed layers of the serous membrane.

(b) PERICARDIAL FRICTION AND ATTRITION.

Friction is the mechanical result of the heart's movements in contact with a roughened membrane. It occurs at the first stage of every pericarditis, whether subsequently adhesive, attritive, or exudative. Continuous friction will wear itself out after varying periods, the pericardial surface becoming once more polished. So long as roughness of the surface persists it is apt to give rise to audible friction-sounds.

FRICTION SOUNDS AND FRICTION MURMURS.

Pericardial friction may be :
>   (a) External, or
>   (b) Internal, to the sac.

(a) EXTERNAL PERICARDIAL FRICTION SOUND.

By this is meant the friction sound set up by the heart's action between an inflamed pleuro-pericardial membrane and an inflamed pleural surface (be it parietal or visceral), the inner surface of the pericardial sac remaining unaffected. This friction is always difficult to tell from internal pericardial friction.

When it occurs over the anterior, or subchondral, surface of the pleuro-pericardial fold its locality is rather more easily traced,

(1) because in that situation friction due to respiratory movement is plainly audible in the interval between the heart-beats ; and

(2) because friction due to the movements of the heart is markedly diminished when the chest is kept at rest in the position of full inspiration.

When, however, friction is limited to the lateral, deep, or pulmonary surface of the pleuro-pericardium, the tests given above may not be so effectual ; and the distant rubbing sound, being conducted through the heart's substance to the præcordium, may appear to be independent of any pleural origin or influence.

(b) INTERNAL PERICARDIAL FRICTION SOUNDS.

These are the sounds special to pericarditis properly so called. We may study with profit :

>   Their audible characters,
>   Their site,
>   Their rhythm.

(a) *Their Audible Characters.*

Superficial character is special to the group ; but the ear distinguishes a coarse and a fine quality of friction. Moreover, the peculiarities of the sound vary much in different cases, and various names are used to express these varieties. The names may be arranged in the two following groups :

Coarse or Rough Friction Sounds :
{
" Coarse rub " ;
" creaking, or new-leather sound " ;
" churning " ;
" cog-wheel sound " ;
" grating " ;
" scraping " ;
" sawing " ;
}

Fine or Soft Friction Sounds :
{
" Soft rub " ;
" scratching " ;
" grazing "
" rustling " ;
" blowing," or " pericardial friction-murmur."
}

An undulation in the intensity of the sound, as though this were advancing and receding, is very characteristic of pericardial friction. It is due to the influence of the respiratory variations in the size of the lungs.

(b) *Their Site.*

Intra-pericardial friction sounds are by nature :

Localised ;
Not carried along the great vessels ;
Influenced by local pressure ;
Not suppressed by holding the breath.

A very common seat for the earliest indications of friction is the midsternal or basic region of the heart. The sound in this case is not conducted over the whole præcordium from the part implicated. Similarly in general pericarditis friction is not heard beyond the præcordium. Pressure, locally applied, intensifies the friction.

(c) *Their Rhythm.*

The Rhythm may be of three kinds :

Single,
Double, or
Three-timed.

(1) A single rub is occasionally heard. Further pressure from the stethoscope will in most of these cases set up a double sound, although sometimes the friction remains single.

(2) The sound commonly described as special to pericardial friction, is the " to-and-fro " or " see-saw " sound.

(3) The truly typical " three-time," or " cantering," sound will be heard in most cases, if carefully listened for. It is apparently due to the successive movements accompanying the auricular systole, the ventricular systole, and the ventricular diastole.

PERICARDITIS WITH EFFUSION.

In this form of the disease there are two stages :

The stage of dry inflammatory fibrinous exudation ;

The stage of fluid effusion.

The duration of the first stage varies ; it is usually short. The fluid effused during the second stage may be

serous,

sero-purulent, or

purulent.

The effusion may be rapidly fatal ; or toleration may be established, and the accumulation may gradually assume considerable proportions. The subsequent history of pericardial effusion belongs to clinical medicine.

It is noteworthy that pericardial friction does not in all cases cease to be perceived after the occurrence of effusion. The persistence of friction is probably connected with the position assumed by the heart as a result of the distension of its serous sac. Fig. LIX. illustrates the tendency of the heart to become more median in position at the same time as its apex is lowered.

" Dropsy of the Pericardium,"

" Hydrops Pericardii,"

" Hydro-pericardium "

are equivalent terms. If used without further qualification, they convey the idea of " passive effusion " (a more explicit and preferable synonym).

A passive effusion may be in its onset acute or chronic. Therefore, unless when specially stated, neither " Pericardial dropsy " nor " Hydrops pericardii " nor " Hydro-pericardium " imply the existence of acute pericarditis.

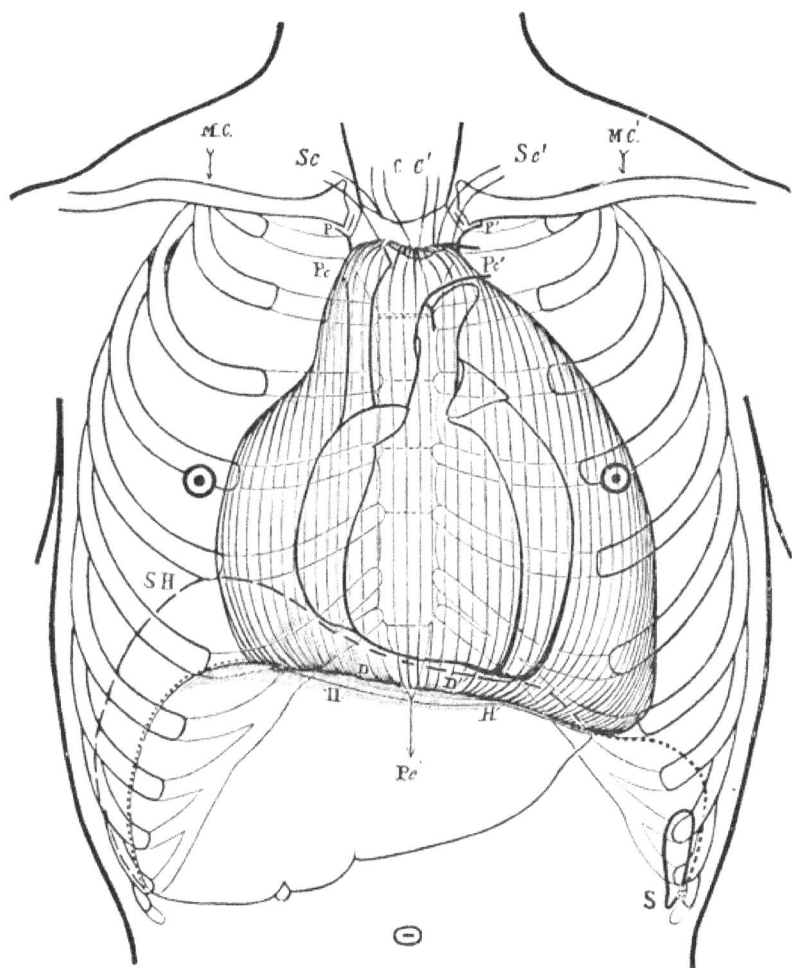

Changes in the axis and in the level of the heart in pericardial effusion. DD' the diaphragm, and HH' the liver depressed to the level of the tip of the xiphoid. The cardiac apex is lowered and brought nearer the middle line than normal.

PRACTICAL POINTS IN THE AUSCULTATORY DIAGNOSIS OF PERICARDITIS.

In all cases of audible pericardial friction two questions must be considered :

(1) Is the sound endocardial or exocardial ?

(2) If not endocardial, is the sound due to internal pericarditis or to external pericarditis (pleuro-pericarditis) ?

I.

DIAGNOSIS BETWEEN ENDOCARDIAL AND EXOCARDIAL MURMURS.

Sometimes the sound of pericardial friction closely resembles a blowing sound. This variety of friction sound has been described as a " Friction Murmur," or " Exocardial Murmur," in contradistinction with the endocardial or true heart-murmurs. Owing to this similarity diagnosis is often very difficult. Attention to the following points may be of service in determining the nature of the doubtful sound.

(1) Since friction frequently results from a limited roughness, the locality, if it should happen to be one unlikely for an endocardial murmur, may help the diagnosis.

(2) The conduction of endocardial murmurs takes place in definite directions, and in the case of some of them, along the great vessels. In pericardial disease diffusion of the sound either does not occur or it does not follow the directions special to valvular murmurs.

(3) The time at which the exocardial murmurs are heard does not occupy the same relation to the time of the normal heart-sounds as would belong to the endocardial murmurs.

(4) If the pressure of the stethoscope be increased, an endocardial murmur would probaby not be altered ; whereas a marked increase may be observed in the loudness of an exocardial murmur.

(5) Lastly, if the patient's posture be changed, an exocardial murmur would be more apt to change than an endocardial murmur, although posture frequently modifies the intensity of a cardiac bruit.

II.

DIAGNOSIS BETWEEN INTERNAL AND EXTERNAL PERICARDIAL FRICTION
SOUNDS.

Does the rough layer of fibrin, conveying friction sound to the ear or
thrill to the hand, line the pericardial sac ; or does it line the pleural cavity,
where the latter overlies the heart ?

(1) A conclusive answer may be obtained by interrupting the movements
of respiration. If the sound should cease as soon as the breath is
stopped, this will clear up all doubt. If, however, friction should
persist, its source is still open to question ; for an uncomplicated
external pericarditis may occasion a rub synchronous with the
heart's action and independent of any respiratory movement,
although produced within the pleura.

(2) If the rhythm be distinctly cantering, and the sound made up of
auricular systolic, ventricular systolic, and diastolic rubs, the seat
is almost certainly intrapericardial.

(3) In the absence of the typical cantering sound, diagnosis may be
assisted by modifications of the sound which may arise when the
breath is held, whether in inspiration or in expiration, and especially
by those which may be brought about by exaggerated respiratory
movements.

In a proportion of cases the diagnosis cannot be made by unaided
auscultation.

PART VI.

PRACTICAL ILLUSTRATIONS OF THE METHOD OF USING DIAGRAMS AND SYMBOLS FOR THE RECORD OF PHYSICAL EXAMINATIONS OF THE HEART.

ON THE VALUE OF NEGATIVE NOTES.

The value of negative notes of symptoms and physical signs is appreciated by all clinical workers of experience. It is not enough to have ascertained the absence of a physical sign ; the fact must be recorded at the time, or it is lost to the reader of the notes, and may ultimately be forgotten by the observer himself.

The excellence of negative notes varies with the degree in which they are circumstantial. A typical negative note addresses itself to a single and definite fact. Notes such as " heart healthy " or " no cardiac murmur" may be described as "collective" negative notes. They are valuable, but their value is inferior to that of the circumstantial note, in which may be traced each separate examination tending towards the collective result. The best note is that which bears evidence that each valve has been examined and found healthy. A note such as the following :

" Heart examined :
no onward or regurgitant aortic murmur ;
no onward or regurgitant mitral murmur ;
no onward or regurgitant tricuspid murmur ;
no onward or regurgitant pulmonary murmur,"

carries conviction as a statement of fact, but would be a serious addition to the labour of each examination.

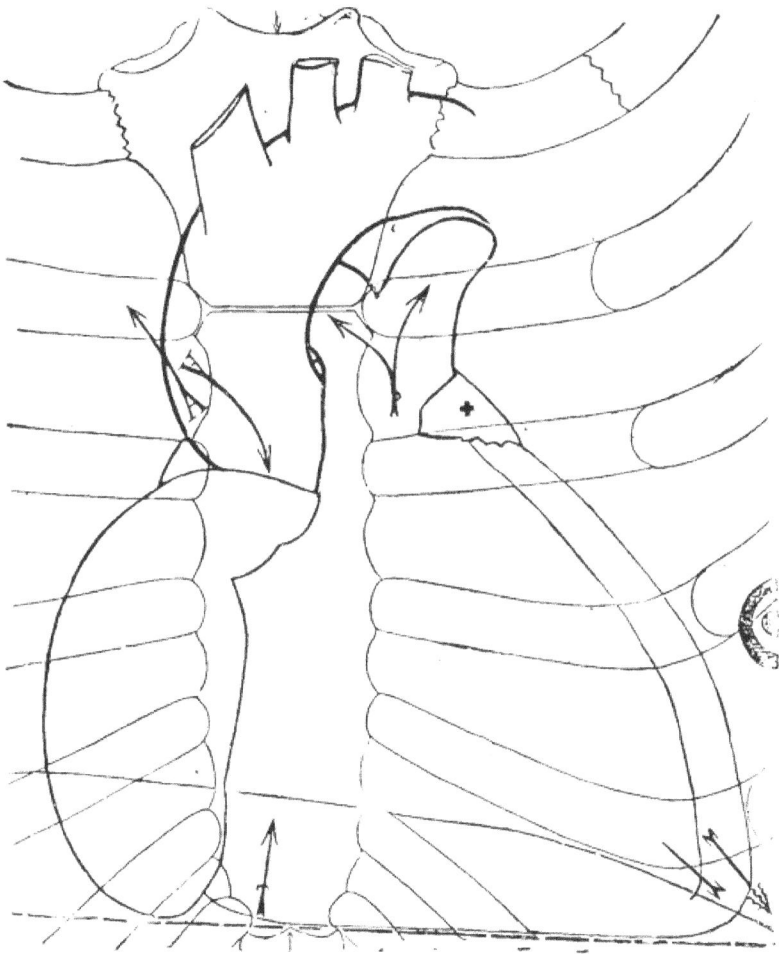

FIG. LX.

THE PREPARED CLINICAL OUTLINE BEFORE USE.

The arrows represent the valvular murmurs (two of which have been omitted as rare). The cross refers to Naunyn and Balfour's murmur. *The arrows are to be struck out if the murmurs be absent.* The heart sounds are supposed to be normal; if otherwise, their loudness, or reduplication, etc., may be expressed by symbols. The heart boundaries, depicted as normal, may need to be altered in accordance with the results of percussion.

The graphic method supplies us with an easy method of fulfilling the indication without any unfair tax on the observer's pen.

THE PREPARED CLINICAL OUTLINE.

If a suitable diagram bearing the symbols of all the likely murmurs be used in recording each physical examination, the absence of each murmur, as soon as ascertained by examination, could be recorded by striking out the corresponding symbol, which, in the case of a murmur found to be present, would be allowed to stand.   Negative notes thus taken would be circumstantial, yet they would occupy neither much time nor much space, the diagram remaining available for any additional graphic or other note bearing upon the percussion of the heart, or the quality of the heart sounds, or the extent of the area of conduction of any murmurs present, etc.

Whilst it contributes to the rapidity and to the definiteness of negative notes, the Prepared Clinical Outline promotes systematic examinations, and serves to remind us of the various points for investigation.   This feature renders this method of notation specially useful for the student who has to acquire the habit of completeness in cardiac examination, and also to the busy practitioner and to the investigator who should not depart from it.

To the practitioner and to the special worker alike the use of serial graphic records is a great help in difficult and complex observations such as cardiac disease necessitates.   A repetition of the same chart for each successive examination permits progressive changes to be accurately followed and sudden changes to be timed with precision.

Let us take as an instance a case of Rheumatic Fever with threatened endocarditis, in which freedom from any valvular murmur had existed until a given day, when a murmur was first plainly heard.   The use of the diagram, as part, as it were, of each physical examination, would have ensured a faithful daily entry of the hitherto negative results, and the case would stand as a complete and ready record of the cardiac events.

In most rheumatic cases, however, and in others also, the change is usually not an abrupt one.   The heart-sounds will have altered their character before being replaced by murmurs, and the change will probably have first been observed over definite areas of the præcordium ; again the rhythm of the sounds may have undergone some alteration ; and the dulness

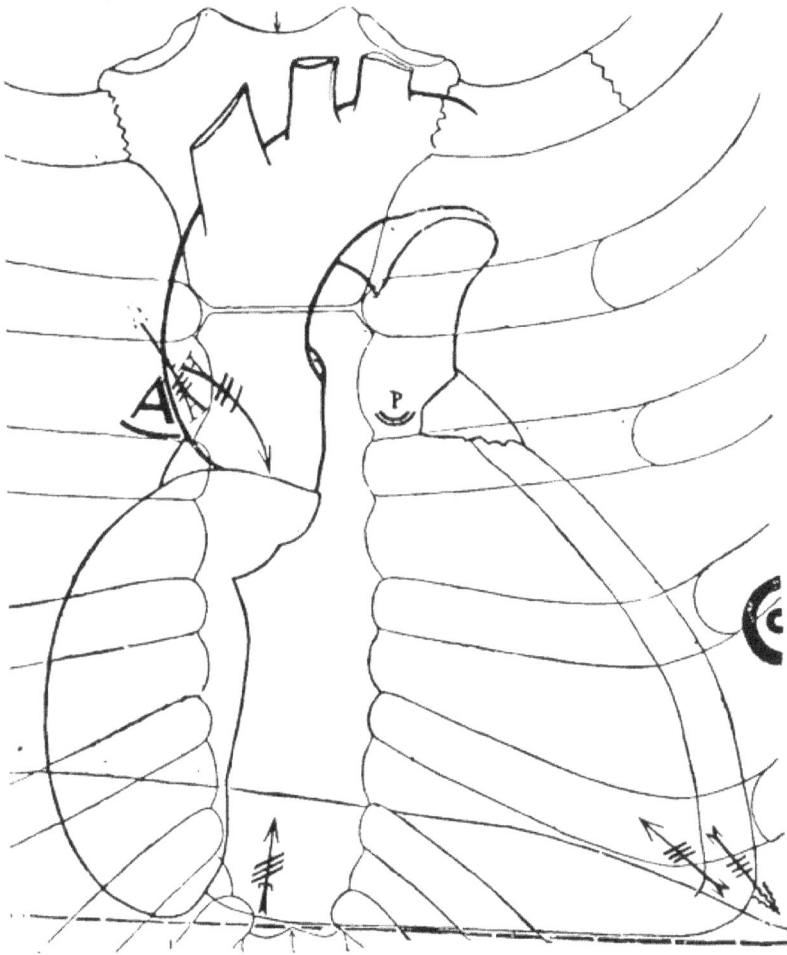

FIG. LXI.

THE PREPARED CLINICAL OUTLINE IN USE.

All the arrows having been struck out, the heart is shewn to be free from murmurs. The large symbol in the right second interspace indicates considerable loudness of the aortic sound. The smaller symbol in the left second interspace means that a reduplication is heard at the site of the pulmonary second sound.

FIG. LXII.

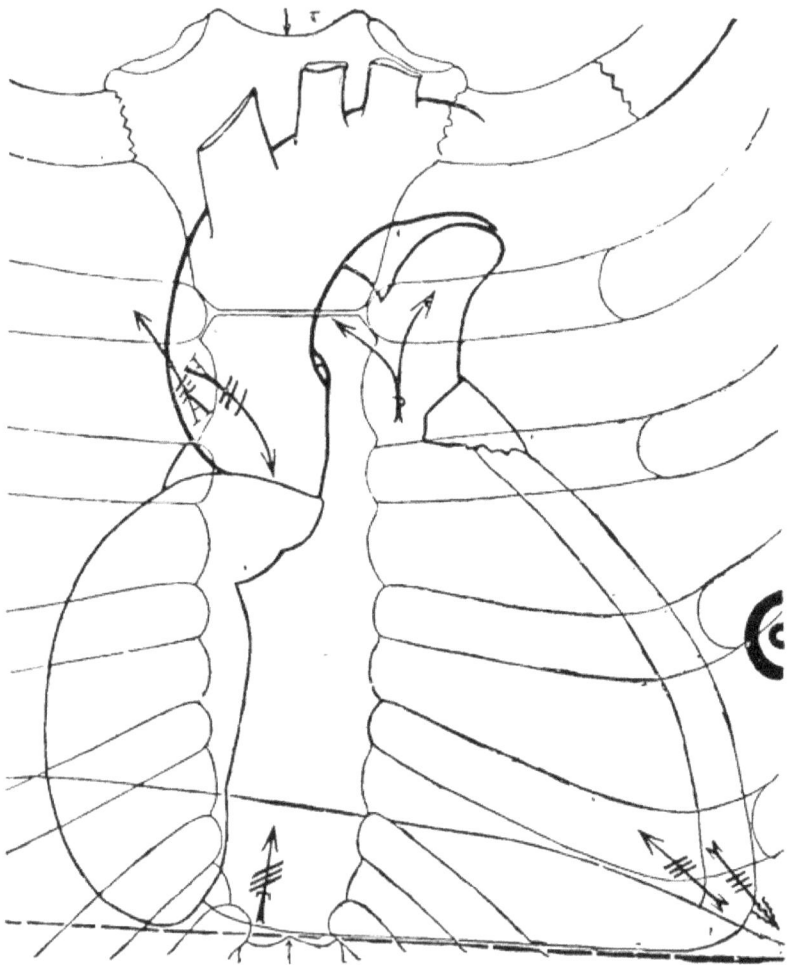

THE PREPARED CLINICAL OUTLINE IN USE.

Five arrows having been struck out, as well as the cross, the corresponding murmurs are absent. One arrow remains, that which represents the
systolic pulmonary murmur. The Outline does not shew whether the
murmur was thought to be functional or organic. A note stating the
observer's opinion should have been added to the record.

may have increased towards the left or towards the right. A graphic nota-tion of facts of this kind will not be accurate and intelligible unless fair space be allotted to it. This is the reason for the size of the Outline which otherwise might have been thought inconveniently large. Size is specially important in recording the relations of the heart to the ribs and interspaces, and its comparative dimensions as found by percussion.

Whenever accuracy and completeness of the record are not essential, or when a single object, be it a murmur, an abnormal sound, an abnormal impulse or an abnormal dulness, merely requires to be noted, the Elementary Clinical Outline, Fig. I., will be found easy to work with in proportion to its simplicity.

Reverting to the Prepared Clinical Outline, we see in Fig. LXI. an illus-tration of its application to the recording of the absence of murmurs. Fig. LXII. shews a positive record of a murmur, whilst other murmurs are shewn to be absent. Lastly the use which may be made of the symbols for heart-sounds is illustrated in Fig. LXI.

THE END.

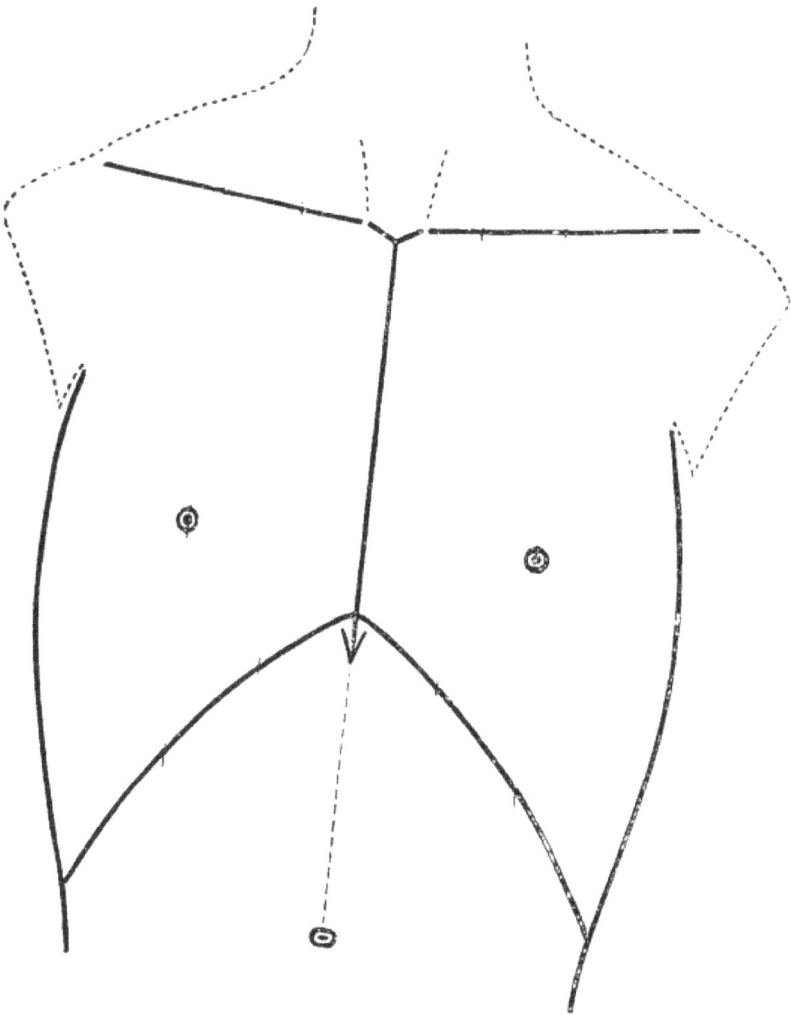

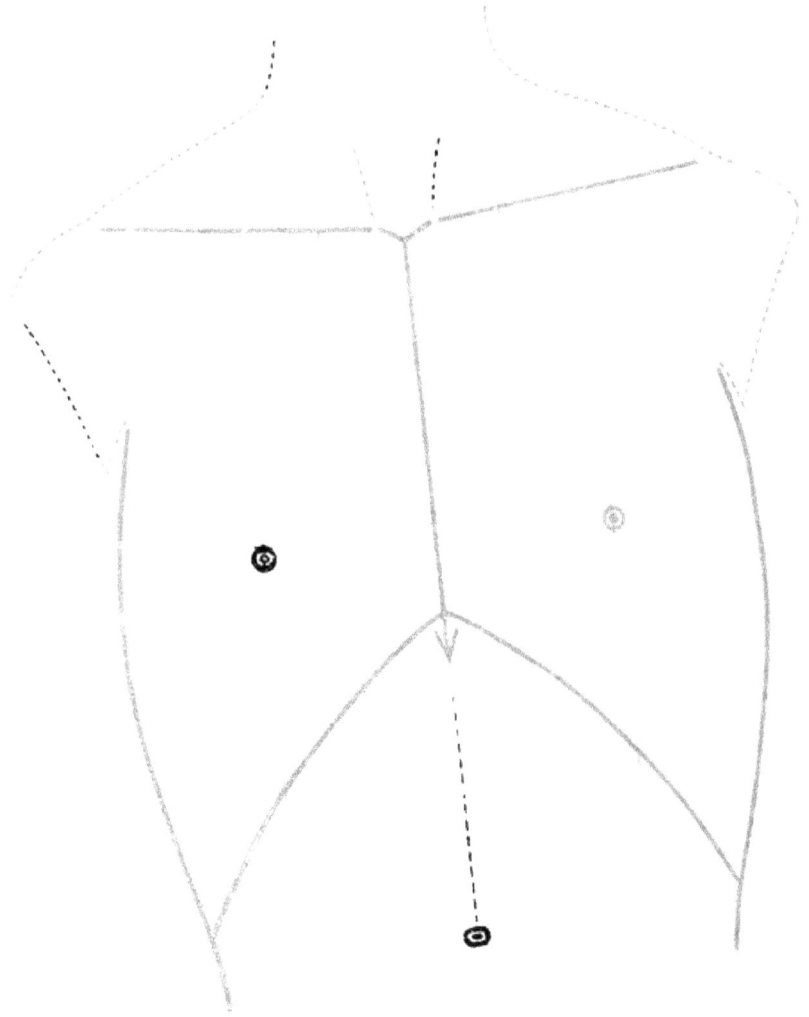

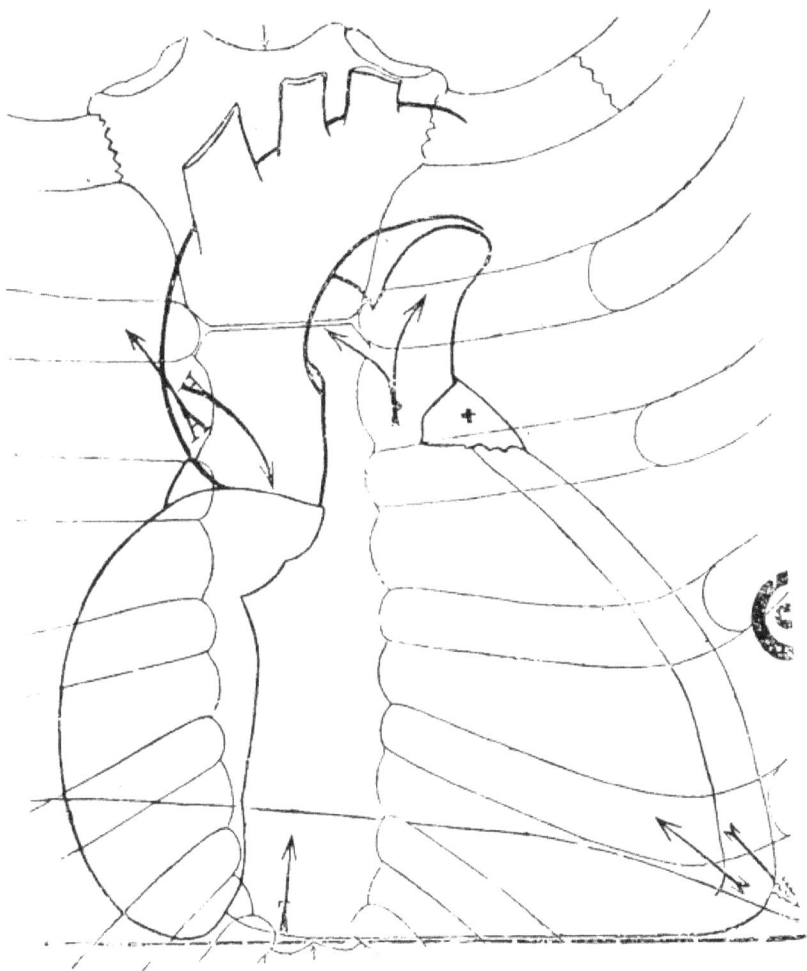

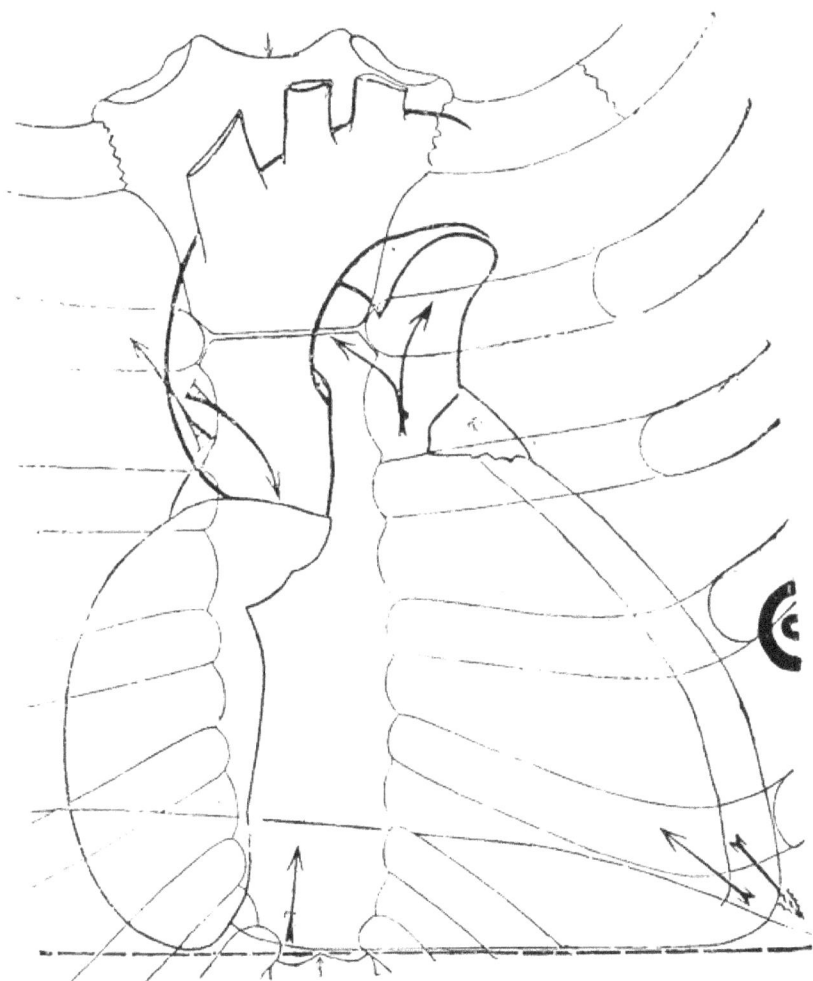

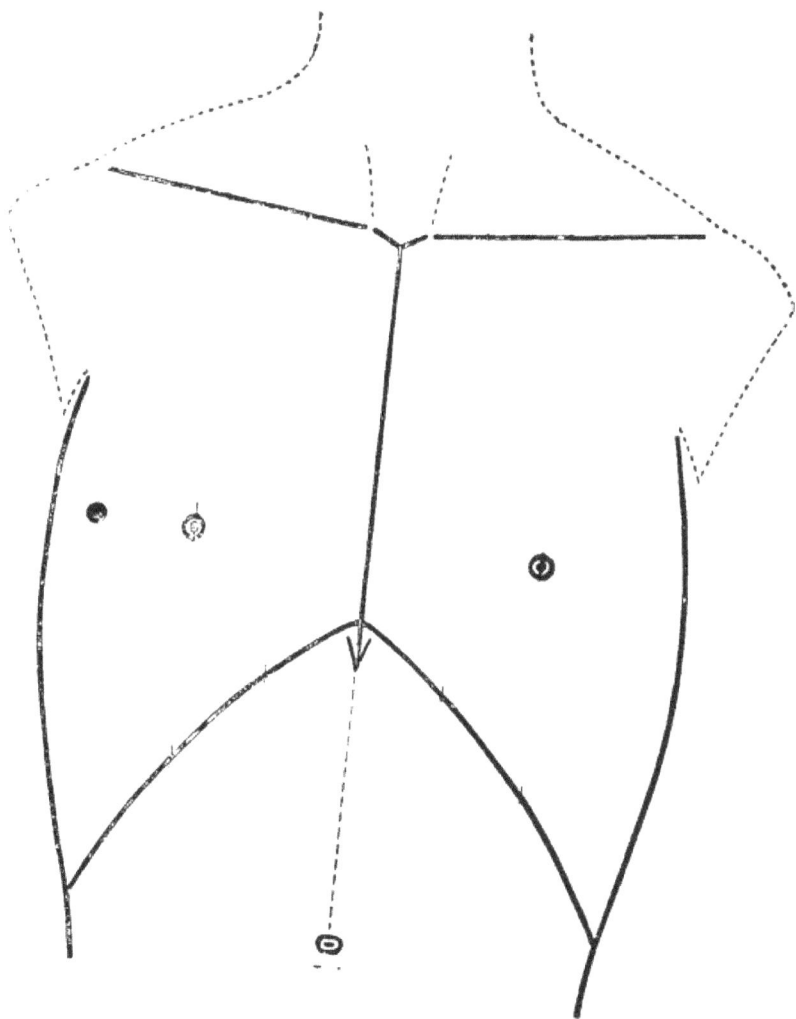

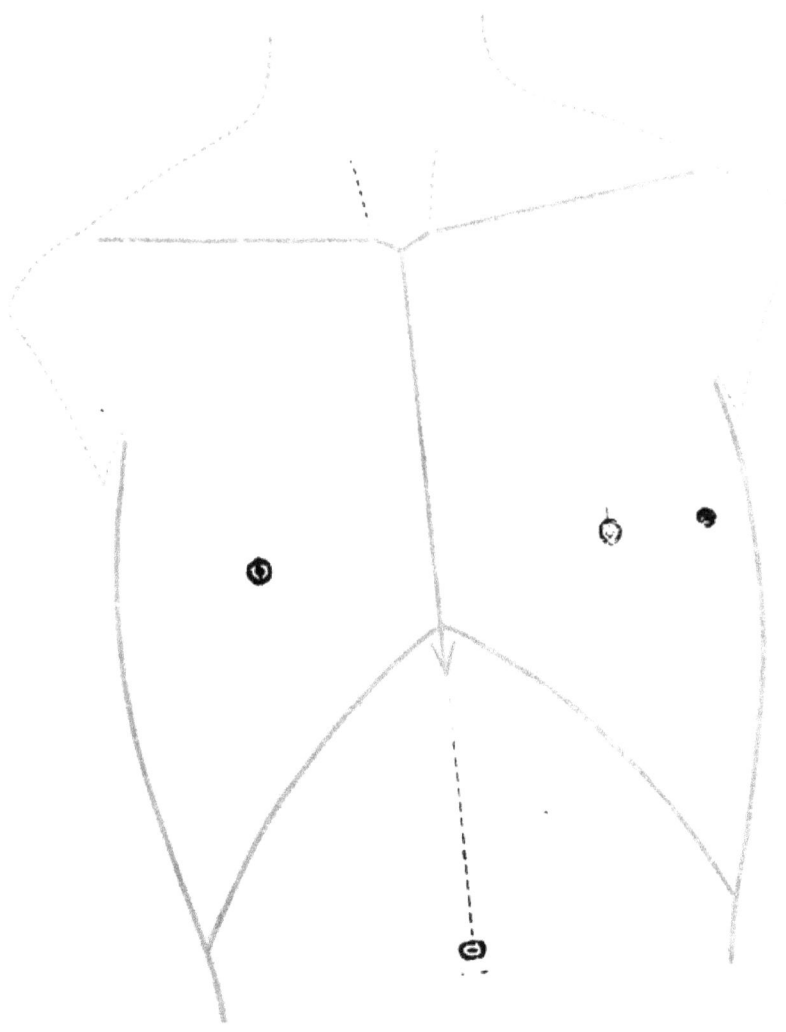

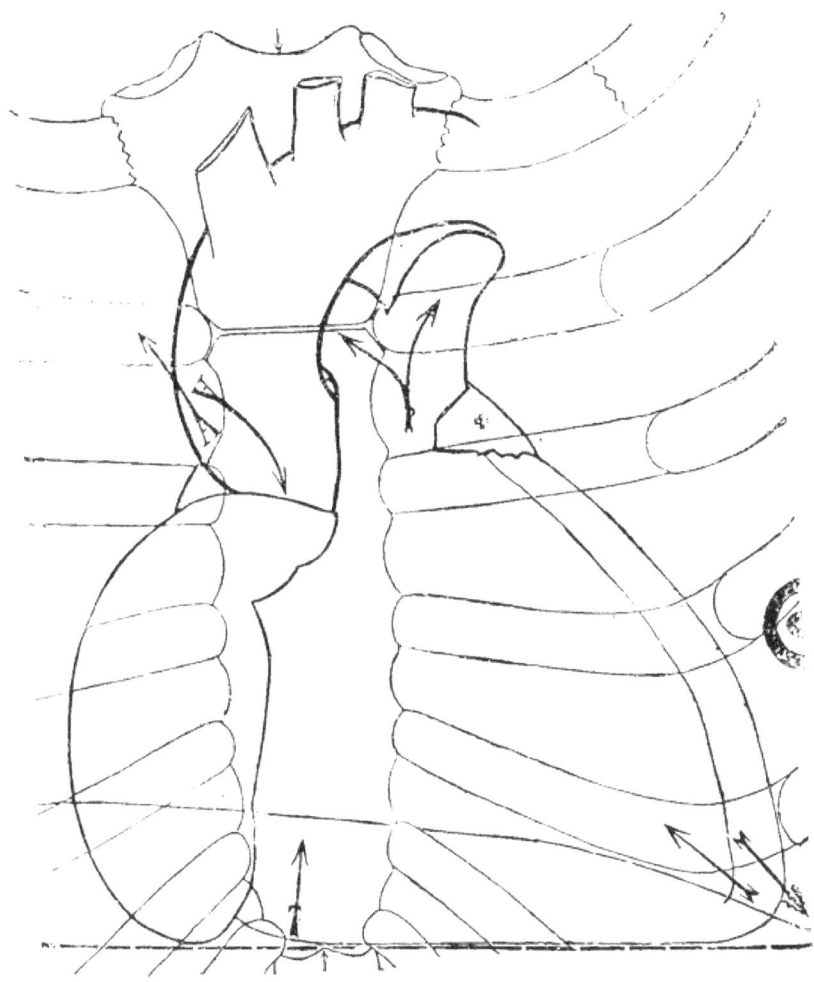

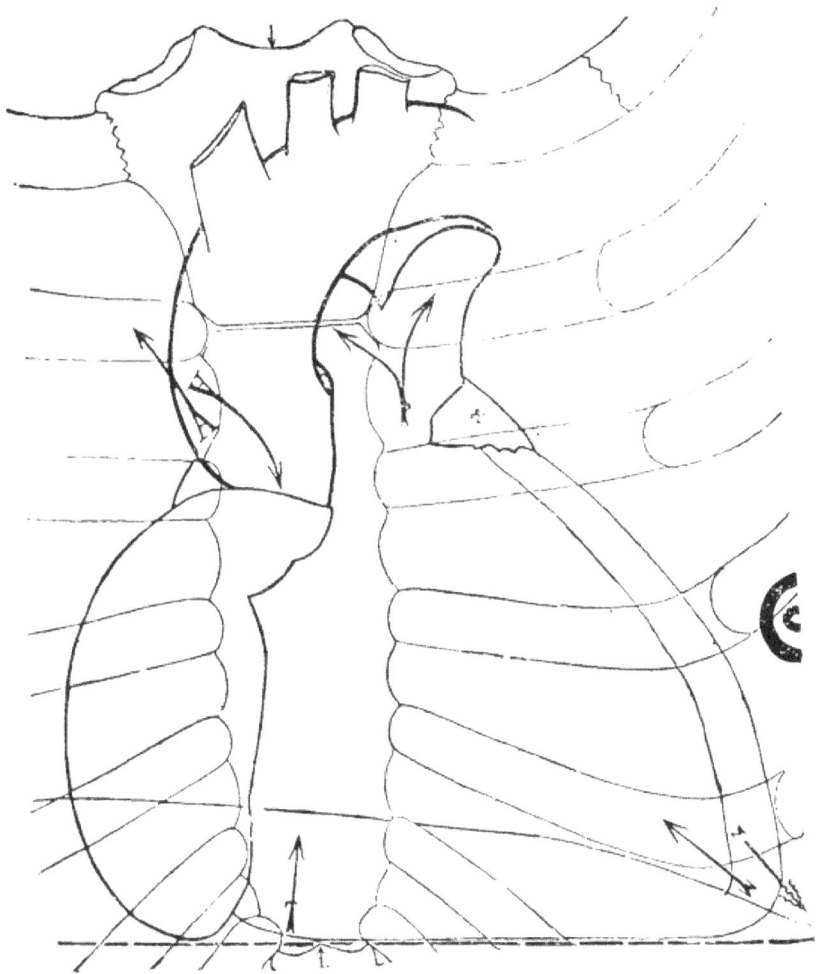

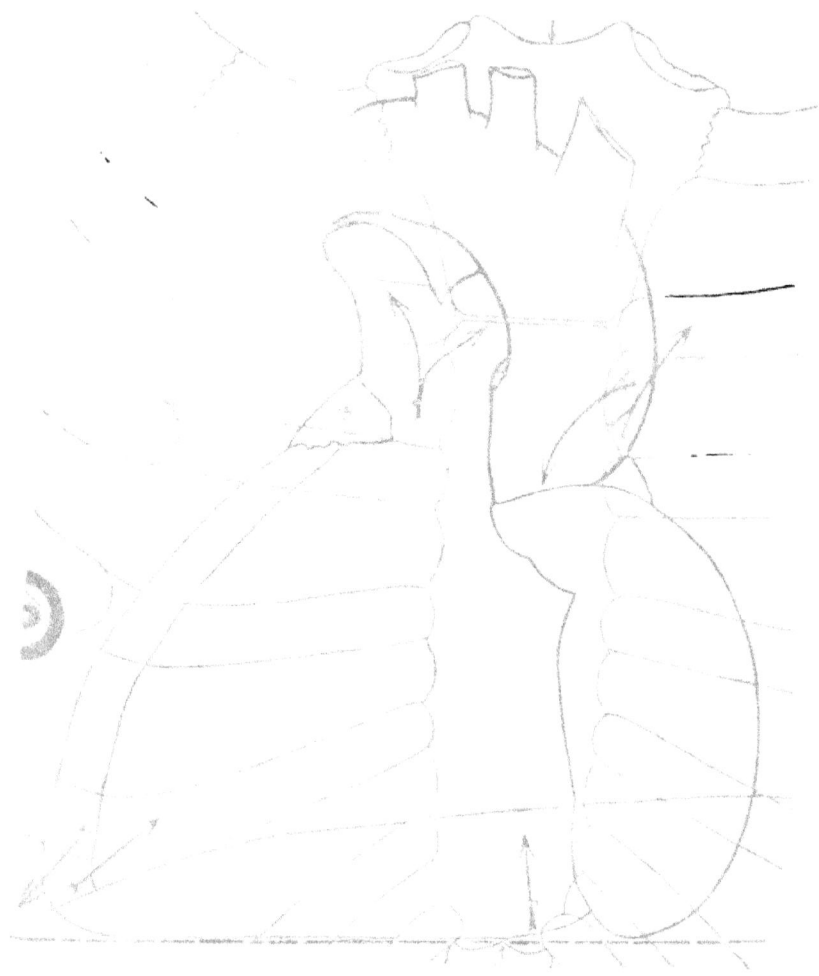

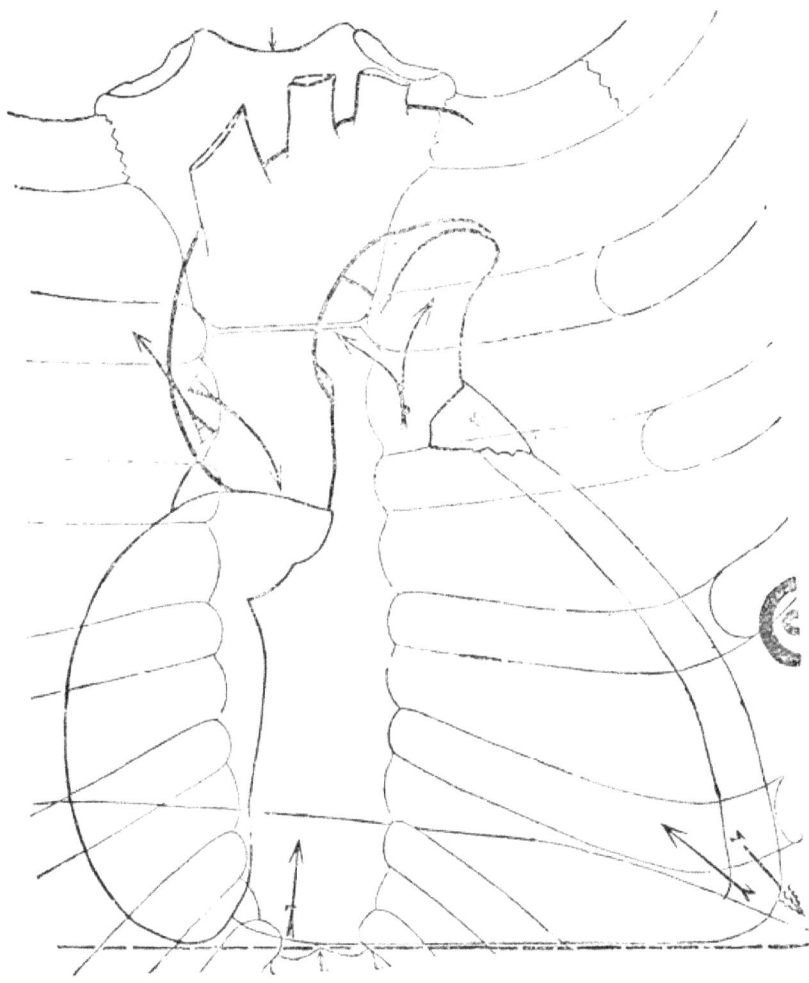

www.ingramcontent.com/pod-product-compliance
Lightning Source LLC
Chambersburg PA
CBHW030844270326
41928CB00007B/1204